SLOW BURN

Slow Burn

THE HIDDEN COSTS OF
A WARMING WORLD

R. JISUNG PARK

PRINCETON UNIVERSITY PRESS

PRINCETON & OXFORD

Copyright © 2024 by Princeton University Press

Princeton University Press is committed to the protection of copyright and the intellectual property our authors entrust to us. Copyright promotes the progress and integrity of knowledge created by humans. By engaging with an authorized copy of this work, you are supporting creators and the global exchange of ideas. As this work is protected by copyright, any reproduction or distribution of it in any form for any purpose requires permission; permission requests should be sent to permissions@press.princeton.edu. Ingestion of any IP for any AI purposes is strictly prohibited.

Published by Princeton University Press
41 William Street, Princeton, New Jersey 08540
99 Banbury Road, Oxford OX2 6JX

press.princeton.edu

GPSR Authorized Representative: Easy Access System Europe - Mustamäe tee 50, 10621 Tallinn, Estonia, gpsr.requests@easproject.com

All Rights Reserved

First paperback printing, 2025
Paperback ISBN 9780691224183

The Library of Congress has cataloged the cloth edition of this book as follows:

Names: Park, R. Jisung, 1986– author.
Title: Slow burn : the hidden costs of a warming world / R. Jisung Park.
Description: 1st ed. | Princeton : Princeton University Press, [2024] |
    Includes bibliographical references and index.
Identifiers: LCCN 2023041377 (print) | LCCN 2023041378 (ebook) | ISBN
    9780691221038 (hardback) | ISBN 9780691221045 (ebook)
Subjects: LCSH: Environmental economics. | Climatic changes—Economic
    aspects. | Climatic changes—Social aspects. | BISAC: BUSINESS &
    ECONOMICS / Environmental Economics | SCIENCE / Global Warming &
    Climate Change
Classification: LCC HC79.E5 P3478 2024 (print) | LCC HC79.E5 (ebook) |
    DDC 363.7—dc23/eng/20241103
LC record available at https://lccn.loc.gov/2023041377
LC ebook record available at https://lccn.loc.gov/2023041378

Editorial: Joe Jackson, Morgan Spehar, Emma Waugh
Jacket/Cover: Katie Osborne
Production Editorial: Elizabeth Byrd
Production: Erin Suydam
Publicity: James Schneider, Kate Farquhar-Thomson
Copyeditor: Crystal Atamian

Jacket/Cover Credit: Melani / Adobe Stock; Life of Pix / Pexels

This book has been composed in Arno

To my wife, Young Ji, whose wholeheartedness
is my inspiration and joy.

And to our parents, who gave their all to give us wings.

CONTENTS

Acknowledgments   ix

Introduction   1

PART I   23

1   Thinking Fast and Slow about Climate Change   25

2   Physical versus Human Capital   34

3   Where There Is Smoke   53

PART II   65

4   Anecdotes, Data, and the Problem of Causality   67

5   How Heat Hurts   79

6   Temperature and the Wealth of Nations   99

7   Peace and Peace of Mind in a Hotter World   126

PART III — 149

8 Climate Change and Economic Opportunity — 153

9 Climate Inequality Close to Home — 177

10 The Hidden Determinants of Climate Vulnerability — 196

PART IV — 213

11 Never Too Late — 215

12 Beyond Silver Bullets — 241

Conclusion: The Rest of Creation — 277

Notes 283

Index 317

ACKNOWLEDGMENTS

THIS BOOK would not have been possible without the engagement and feedback provided by gifted friends and mentors who have encouraged, challenged, and guided me.

I would like to thank Michael Chwe, Abdul El-Sayed, and Chad Zimmerman for planting and working the seeds of the initial idea, as well as Murray Biggs, Wing-Kit Chu, James Hacker, Scott Moore, Caroline Park, Jun Park, Jennifer Wilson, and Taylor Yates for providing helpful feedback on early drafts. I am grateful to Li Chen for helping me convey the book's essential messages in my own voice, and to Patrick Callahan for his influence in helping to retain its moral integrity.

I am deeply grateful to coauthors and collaborators in my research, including Alan Barreca, A. Patrick Behrer, Mark Curtis, Joshua Goodman, Lucas Husted, Isaac Opper, Nora Pankratz, Paul Stainier, and Anna Stansbury, without whom the research that forms the bedrock of this perspective would not have been possible.

My thinking on this subject has benefited enormously from conversations with innumerable colleagues and mentors, including especially Joe Aldy, Marshall Burke, Raj Chetty, JR Deshazo, Geoffrey Heal, Cameron Hepburn, Matthew Kahn, Larry Katz, Nat Keohane, Matt Kotchen, Frances Moore, Matt Neidell, Jeffery Sachs, Jeff Shrader, Robert Stavins, and Gernot Wagner.

This book also benefited greatly from excellent research assistance by Ina Drouven, Michael Ma, Paul Stainier, and Jonah Stewart.

Finally, I am indebted to my editor, Joe Jackson, who exhibited consummate professionalism and provided thoughtful feedback throughout the project, as well as the many dedicated members of the Princeton University Press team.

Introduction

WE BEGIN with two tales from antiquity.

You may be familiar with the story of Mount Vesuvius, whose eruption covered an entire city in lava and ash. In the year A.D. 79, a long-dormant volcano on the southern coast of Italy suddenly detonated, sending a colossal plume of ash and smoke into the sky. The ensuing eruption wreaked havoc on the nearby city of Pompeii as scalding debris rained down with ferocious intensity, at a rate later estimated to have reached over 1.5 million tons per second.[1]

Some residents tried to evacuate the city. Others sought refuge in whatever shelter they could find. A second, more violent eruption is thought to have sent a surge of pyroclastic material across the coastline, likely leading to the nearly instant demise of those who had stayed behind. Researchers estimate that in the immediate aftermath of the blast temperatures in the dwellings of Pompeii may have reached 300°C, which would be enough to kill those sheltering inside in mere seconds.[2]

For the city of Pompeii, the eruption of Mount Vesuvius was, by any measure, catastrophic. Buried under a layer of lava and ash several meters deep, the once thriving metropolis was effectively erased from the map, not to be seen again until its excavation over a millenium later.

If starting a book about climate change with a parable of environmental catastrophe feels like a depressingly familiar trope to you, that's probably because it is.

The stories we tell ourselves about climate change too often have a subtext of looming cataclysm, focusing on the disaster scenarios, invoking fear and despair. We are in the eleventh hour, or worse, we have already crossed the irreversible threshold, the point of no return. Or so the narrative often goes.

This book seeks to challenge this tendency, to question the appropriateness of this familiar doomsday narrative. Not because climate change is not a serious problem. As we will see, if anything it may be a more insidious threat to human flourishing than many realize. But because a growing body of evidence suggests that the familiar *climate catastrophe* framing may be missing some of the most important features of the real climate change story. As a consequence, it may be hindering us from thinking more proactively about potential solutions.

Our second tale is about the fall of Rome.

The Western Roman Empire was what many consider to be the first and at the time largest world empire. At its zenith, it stretched across Europe, North Africa, and Western Asia, encompassing much of modern-day Italy, Spain, France, England, Morocco, Greece, and Turkey. Rome's military conquest and cultural accomplishments are legendary. Echoes of its extensive road network and extraordinary architecture endure to this day.

Economic data from that period is scarce at best, but archaeological evidence suggests that denizens of the Roman Empire may have experienced some of the highest material standards of living in the pre-industrial era.[3]

While historians debate the precise causes of the Roman Empire's downfall, one thing seems clear. The fall of Rome was a gradual decline born both of internal and external forces, a slow deterioration stretched out over decades if not centuries.

We may never know the precise combination of causes that precipitated its decline. Theories include the so-called barbarians at the gates hypothesis, which points to the growing frequency and intensity of violent invasions that chipped away at military defenses and financial stability; disease-driven decline, in particular the Antonine Plague of the late second century, which may have precipitated an erosion of health and human capital; political and institutional factors, including infighting among Rome's ruling classes; as well as rising economic inequality and social unrest, which some believe may have led to diminished political participation and a flight away from urban centers by the elite.[4]

In part because the decline was gradual, scholars continue to debate the precise date at which the Roman Empire "ended." Some would argue that the commonly accepted date of A.D. 476, when a coup overthrew emperor Romulus Augustulus, may mischaracterize the reality that spheres of Roman rule and influence persisted in many parts of the world for centuries thereafter.

What seems less debatable is the fact that gradually the Roman Empire's economic and cultural grandeur lost its sheen, as once mighty cities saw their populations decline and thriving interregional trade slowed to a crawl, leading to lower standards of living, diminished military power, and faded cultural and political influence overall.[5]

Slow Burn

This book is about the deeper consequences of a hotter planet. It isn't a climate change horror story. Nor is it a contrarian account of why we should all relax and direct our attention elsewhere.

It is an invitation to view the climate problem through a slightly different lens. One that is less about headline-grabbing catastrophes and more about the slow burn—the largely invisible costs that may not raise the same alarm, but which, in their pervasiveness and inequality, may be much more harmful than commonly realized, and call for swift action in ways you might not expect.

The central premise of this book is that the subtle setbacks of a changing climate may comprise some of its most important challenges: imperceptibly elevated health risks spread across billions of people; pennies off the dollar of corporate profitability; the immobilizing erosion of coastal and agricultural livelihoods; young people learning less, old people remembering less, many of us arguing more.

This isn't to say that the risk of the planet's catastrophic heart failure is unimportant, but rather to suggest that the near certainty of its chronic inflammation may be reason enough to act, especially if the ripple effects of such metaphorical inflammation prove systemic in unexpected ways, and given that such chronic inflammation may, for some, prove fatal.

In this book, I highlight how our intuitions may be ill-suited for making informed decisions about climate change. Given the layer cake of uncertainty that climate change presents, our minds may be especially prone to defaulting into incomplete heuristics that paint a fatalistic picture of black and white, despite the many decision-relevant shades of gray in between.

This is not merely an academic distinction, especially when one considers the fact that getting climate policy right is not a one-time, all-or-nothing decision. Its massive scale and complex political economy mean that it will likely require sustained policy engagement and private sector investment over the course of several decades, not to mention continual balancing of salient current costs and murkier, often nonmonetary, future benefits. Moreover, its hidden and often heterogeneous social impacts demand a more nuanced understanding of climate vulnerability and adaptation, particularly for the world's poor, so that they may be acted upon swiftly and in an evidence-based way.

Both factors point to a real need for a balanced, data-driven understanding of the issue. As we will see, many of the effects of climate change may already be affecting our pocketbooks and our quality of life in not so obvious ways.

The Hidden Costs of a Warming World

This book attempts a stylized synthesis of a new wave of economic research on this topic. It draws from carefully conducted studies at the cutting edge of modern empirical social science, particularly those that bring real-world data and careful disentangling of cause and effect to bear on what historically has been a modeling and assumption-heavy enterprise. Many of these studies have only surfaced in the past decade or so.

INTRODUCTION 5

This book is written with an understanding that, for most readers, statistics and econometrics may not have them jumping out of bed in the morning. Through a mix of stories and academic studies, we will unpack what the data tells us about how climate already affects our individual daily lives, in ways both seen and unseen.

For instance, did you know that, depending on the tasks one is responsible for, going to work on a day when the temperature is above 90°F may lead to a 5 or even 50 percent increase in the likelihood of serious injury compared to a cooler day in the 50s or 60s—mostly due to ostensibly unrelated accidents like falling off a ladder or mistakes while operating heavy machinery? Or that, when a student takes an exam on a 90-degree day in a building without working air-conditioning, their performance may fall by up to 10 percent, and that hotter temperature in the classroom may be widening academic achievement gaps between rich and poor, black and white, both in the United States and elsewhere? How about the fact that the number of gunshots in your neighborhood changes on a hot day, and that the number of suicides changes too? Could you guess in which direction?

Suppose that the quarterly earnings of your favorite equity investment already depend on the number of floods in China, heat waves in India, or cold spells in Canada? Or that the adverse effects of a seemingly minor climate event can ripple out across the global supply chain? Would you think differently about the risks posed by climate change?

An empirically nuanced understanding of how climate change affects the many aspects of our day-to-day lives may be a critical input to important decisions: both regarding the ideal stringency of climate mitigation—how quickly we should move away from fossil fuels and at what costs—as well as the adaptation investments necessary to prepare for the warming that is in store.

There are at least two reasons why. First, the costs of the slow burn may in aggregate be larger than the headline-grabbing climate disasters. I realize this depends on what we mean by climate disaster, and may strike some as a bold claim, especially given the doomsday imagery— whether of inundated cities or decimated ecosystems—often associated with climate change. We will examine the data supporting such a notion in the chapters ahead. Provided the evidence adds up, this would suggest a more dispassionate, more immediate, and perhaps less morally charged reason to support mitigating greenhouse gas emissions aggressively. It also suggests that, even when one incorporates known and unknown risk into the equation, "solving climate change" will be more about choosing contributions on a sliding scale than a binary yes/no decision, which implies that, in a very real sense, and contrary to the doomsday slogans, we are never "too late" to make a difference.

Another way this perspective informs our decision-making is that, by looking differently at everyday phenomena we thought we understood, we can begin to appreciate the subtle social disparities that arise in a warmer world. This may influence our view of the seriousness of the climate problem globally. It may also reveal more immediate interventions locally, particularly when it comes to actions needed to prepare for the warming that is already baked in, especially given how much the intensity of climate damage is influenced by human choices.

A proactive stance toward adaptation and resilience may be useful from the standpoint of safeguarding one's own physical and financial security, whether as a homeowner or the head of a Fortune 500 company. It may be vital for ensuring that the ladders of economic opportunity are not fraying for those climbing its lower rungs. And not just at the level of "climate refugees from the third world," but also for the people who service your car, deliver your mail, and those who harvest, prepare, and cook the food you eat.

A perhaps controversial claim of this book is that becoming more familiar with how climate change hurts us should, if anything, give us a greater sense of hope. Not a pollyannaish hope born of wishful thinking, but an active hope born of clear discernment, a sober assessment of climate vulnerabilities and the socioeconomic details of what can be done to help the world adapt.

Immediacy, Inequality, and Uncertainty

When it comes to climate change, three facts loom large.

The first is that there is a lot of warming in store, coming at us more quickly than many scientists used to believe. In part because the speed at which a given amount of CO_2 translates into climate change may be faster than previously understood, significant warming is no longer only a matter of our children's or our grandchildren's future, but our own as well. For instance, there appears to be a positive feedback loop between rising CO_2 concentrations and warmer, more acidic oceans, which reduces the ocean's ability to remove CO_2 from the atmosphere. This means that older climate models that did not take such feedback loops into account may understate the rapidity of warming.[6]

At the time of this writing, the earth has already warmed by around 1.1°C. Even with the ambitious pledges of recent international accords, we are on pace to experience 1.9°C to 3°C of warming by the end of the century.

These numbers may at times be hard to make sense of intuitively. On a cold winter morning, a bit of warming may even sound nice. A more experientially anchored alternative may be to count the number of relatively extreme events, like the projected frequency of days above 32.2°C (90°F) per year. This is still an imperfect measure, but at least it might be a little easier to relate to. After all, most have experienced sweltering summer heat.

By this measure, Rome is expected to see a ten-fold increase in such days by 2050–2060, relative to preindustrial averages. Residents of Atlanta are expected to see fifty *additional* such days per year, on top of the twenty or so per year they experienced prior to anthropogenic warming.

As one gets closer to the equator, this number tends to grow. For the residents of Accra, Ghana, Mumbai, India, or Bangkok, Thailand, this number may be closer to 100. That isn't a typo. One hundred *additional* days per year above 32.2°C. In many such places, most homes do not yet have air-conditioning. Physical acclimatization, while effective to a point, likely has its limits.

Regardless of whether we are able to reduce emissions rapidly starting today, some level of warming is essentially locked in, at least for the next several decades.

This isn't to suggest that reducing emissions won't help. Quite the contrary, especially in terms of reining in warming during the latter half of the century and beyond, reducing emissions will be crucial.

In fact, the somber backdrop of most climate conversations often obscures the reality that, actually, a great deal of progress has already been made. Even as of the early 2010s, many models had humanity on pace to exceed 4°C or more of warming by the end of the century. Middle-of-the-road scenarios now put us closer to 2.5°C by 2100.[7] In 2022, the United States, which is the largest historical emitter and long a laggard when it came to binding climate action at the national level, passed major climate legislation through Congress, which would put it on track to nearly halve US emissions by 2030 relative to 2005 levels.

These are no small achievements. But they may merely be down payments on the century-long mortgage that is climate mitigation: a mortgage whose monthly installments may need to ratchet up over time as the incremental costs of cutting additional emissions grows.[8]

One cannot avoid the fact that, for the time being, there is too much momentum in the system to avoid a significant amount of warming within our lifetimes. The earth's climate is a slow-moving Leviathan, and we've been aggravating it for over a century. As we will see, even if we are able to stave off planetary collapse in the long term, this inertial warming may have very serious—in many cases life or death—consequences for millions of people over the next several decades.

Which brings us to fact number two: inequality. In many parts of the world, this warming comes amid a backdrop of significant and growing economic inequality, particularly between individuals with and without access to the infrastructure and skills needed to participate in the modern global marketplace.

While wages for those at the top of the US income distribution have risen rapidly since the 1990s, wages for most—particularly those without a college education—have stagnated. In real terms, that is, accounting for changes in costs of living, incomes for non-college workers have actually declined.

Similar trends have been documented in much of Europe, though not always as pronounced as in the United States, and in parts of Asia as well. Four decades ago in China, the bottom half of workers earned 27 percent of total national income, while the top 1 percent earned about 6 percent, a ratio of roughly four to one. As of 2015, the income shares have become closer to 15 percent and 14 percent, respectively, or a roughly one-to-one ratio.[9] This means that, in China today, the top 1 percent—mostly the wealthy elite of Shanghai, Beijing, and other major metropolises—now earn almost as much as the bottom 50 percent combined, which amounts to more than 700 million people. (For reference, the numbers for the bottom 50 percent and the top 1 percent income shares in the United States are 12 percent and 20 percent, respectively. In more egalitarian France, the bottom half earn 22 percent, the top 1 percent earn 10 percent.)[10]

How do the risks of climate change vary across the income distribution in your home country or city? Could climate change be a force that further increases the gap between the haves and have-nots? What are the social and economic implications if so?

As we will see, emerging research suggests that the consequences of moderate, noncatastrophic warming may be severe, especially for society's disadvantaged. This raises important questions around the *how* of adaptation and the targeting of climate assistance, many of which can benefit greatly from better data, helping policymakers approach adaptation with more of a precision scalpel than a hacksaw. Given fact number one above—the amount of warming that is baked in—these and other related questions will likely need to inform collective decisions around how best to adapt to a warming world.

This is not exactly the form of inequality one may be accustomed to hearing about in the context of *climate justice*. Most will be familiar with the fact that climate change has historically been a problem caused by historical emissions of rich people in the Global North, many already dead, whereas its consequences are being felt by everyone, including young people in poorer countries of the Global South.

This inequality in attribution has been a major reason why rich and poor countries have not seen eye to eye on climate change. To paraphrase the point made by developing countries: You took your turn at the cheap energy well and raced ahead economically. Now that it's our turn, you're telling us to find a cleaner alternative, which is going to slow us down even further. You can't expect us to be happy with that.

This international aspect of climate equity is critical both for ethical and practical reasons. It seems only fair that those who benefited more from the caffeine boost of cheap fossil fuels should pay proportionately more for the cleanup, so to speak.

One claim put forth in this book, however, is that the implications of climate change and energy policy for *interpersonal* inequality—the variation in how lived experiences are impacted across dimensions of race, gender, immigrant status, and class, within rich and poor countries alike—are likely to be important as well. For instance, these differences may prove pivotal in determining whether the political appetite for a long-run clean energy transition can be sustained, especially if higher energy prices begin to squeeze wallets—particularly for the lower half of the income distribution in countries where the steepest initial cuts will be necessary.

Because of the way the modern economy rewards certain skills in the workplace, the workers who bear the brunt of health and productivity impacts of a warmer world may also tend to be those hit hardest by parallel drivers of inequality—forces like automation, globalization, and skill-biased technical change. Some of these occupations are ones we might expect, like agricultural or construction workers. But others, we might not: manufacturing workers, parking lot attendants, food processing workers, janitorial staff, and warehousing workers also appear to be exposed to workplace climate risks.

Similarly, because of the way housing markets sort people based on income, poorer families may be increasingly likely to live in the most climate hazard exposed areas. The market forces giving rise to such patterns of inequality aren't always obvious to spot, and devising effective interventions can be more challenging than common intuition might suggest.

In this book, we will explore a growing body of research that asks how the contours of climate damages may be altered by the many social and economic dimensions of vulnerability, including the way labor markets and real estate markets are organized, and what kinds of social safety nets people have access to. I will try to make the case that, while the task of mobilizing financial support for poorer nations may indeed be an important one, expanding the evidentiary base regarding how best to design and target adaptation assistance may be equally pressing.

Fact number three is uncertainty. We find ourselves in a landscape of pervading uncertainty. This is true with respect to whether truly catastrophic climate change may be averted in our children's lifetimes. It is also true regarding the many practical decisions around how best to adapt to the changes that are occurring already: that is, whether and how we can meaningfully alter the realized impacts of a rapidly warming world.

At the societal level, there is a growing understanding that reducing emissions will be critical to stave off the worst effects of runaway climate change: a growing consensus on *whether* climate mitigation should be a policy imperative. And yet, *when* and *how* and *at what cost* (borne by whom) remain perennially debated. The swift reversal of bans on drilling for oil in the United States in the fallout of the war in Ukraine and rising energy prices is just one of many possible cases in point, as is the debate surrounding whether and when to phase out internal combustion engines from automotive markets. The tension is often magnified in developing countries, where trade-offs between growth and green can be steeper.

At the individual level, decisions around how best to reduce climate risk—whether as a homeowner, an investor, a manager, or a mayor—often remain distressingly opaque. At what point do you consider selling the family home to move to higher ground? How well-insulated is your retirement portfolio against a potential *climate correction*? Do your child's classrooms—or standardized testing sites—have reliable climate control?

A recent survey of CEOs and CFOs of the largest companies in the world found that 86 percent of respondents agreed that addressing climate risk was important.[11] Some of this may be a direct response to increased shareholder interest; the number of climate-related shareholder proposals has increased fourfold since 2011.

At the same time, more than three in four CEOs and CFOs (77 percent) admit their firms are not fully prepared for the adverse financial impact of a changing climate. Indeed, few appear to have a clear sense of which climate risks matter and why. Is it the risk of flooding, wildfire, or extreme heat waves? Where in the supply chain could climate risk be lurking, and what can be done to address it cost-effectively? Will the risk fall on physical assets like factories and warehouses, or human assets like call center workers?

Perhaps it is no surprise then that eight out of ten executives (82 percent) believe that their companies have little to no control over such an impact on their businesses.[12]

As we will see in the pages ahead, having clearer metrics of physical risks—how much hotter will an area become, or how many more wildfires can we expect per year by 2030?—is probably an important start. But in many cases, it may be insufficient. The more we learn about the socioeconomic details of climate vulnerability, the more it becomes clear that physical and economic risk are not one and the same, and often depend on complex institutional and structural factors.

Considering the potential for systemic financial risks—for instance, the triggering of rapid price revisions and instability in the housing market—as well as the potential fiscal consequences, debates over the proper role of government in adapting to climate change are likely to grow in the years ahead. And these debates will arrive at your doorstep even if the most visible physical hazards, whether they be wildfires or floods, do not.

All of this suggests that there is likely individual and collective value in becoming more fluent consumers of the economic facts, the hidden everyday costs of climate change. While this book will not be able to answer these questions definitively, the hope is that it can equip the reader with some tools to tackle them more proactively.

There is one particular form of uncertainty that we need to address first, and that is the uncertainty around whether climate change's importance in our lives hinges on the notion of climate catastrophe.

How We Think and Talk about Climate Change

I don't want you to be hopeful. I want you to panic. I want you to feel fear every day, and I want you to act as if the house is on fire, because it is.

—GRETA THUNBERG

Just because there is a problem doesn't mean that we have to solve it, if the cure is going to be more expensive than the original ailment.

—BJORN LOMBORG

Much of the popular discourse about climate change has revolved around some notion of civilizational calamity. For some, it is increasingly assumed that the threat is existential. For others, such rhetoric serves as ammunition for derision, grist for the culture war mill, in which environmental fearmongering is but one of many ploys to trample individual liberties, or a case of misguided progressive zeal.

Implicit in these positions are two assumptions that are flawed, or at least worth careful reexamination. The first is that climate change isn't a priority unless the risk is truly existential for humanity, full stop. Only when it threatens civilizational collapse does climate change rise to the level of urgent social challenge. Painting dramatic stories of impending doom and an uninhabitable earth are warranted, even if they sometimes stretch the evidence, because that is the only way most human beings will pay attention—so the argument goes. Conversely, from the opposing standpoint, it is assumed sufficient to poke holes in the most extreme disaster scenarios without stopping to assess whether even the more moderate claims would satisfy the cost-benefit criterion for action.

As I hope the following pages make clear, climate change need not clear the bar of existential global risk in order for it to be worth paying attention to. We may well need the Greta Thunbergs of the world to galvanize interest in protecting the planet for posterity. But we may also need a more dispassionate quantification of the harms that climate change may pose today and in the near future—harms that may, on aggregate, be a far cry from imminent civilizational collapse, but may mean serious loss of livelihood or even life for many individuals, businesses, and communities.

The data increasingly suggests that, well before we reach the level of existential crisis for humanity, the ways climate change will make many of our activities incrementally less productive and fulfilling may warrant serious concern. Whether one's own ethical basis for action is economic efficiency or social equity or both, it may well be that a sober assessment of such subtler damages may provide reason enough to act. This is especially true to the extent that our own personal and collective actions can often amplify or mitigate these damages considerably.

To be clear, acting aggressively to mitigate climate change as a form of insurance policy against potential ecological and civilizational collapse is a highly sensible idea. When possible, we take out insurance against low-probability, high-pain scenarios all the time, whether in the form of property insurance or life insurance. One could argue that focusing our attention on the possibility of annihilation has been instrumental in jolting many of us out of our complacency. In fact, much of the progress that has been made to reduce emissions may not have been possible without the concern and mobilization this narrative has triggered, particularly among young people.

So when I extol the virtues of paying attention to the noncatastrophic, slow burn, it is not to denigrate or dismiss the importance of recognizing the very real—albeit highly uncertain and likely very distant—risk of truly catastrophic climate change. In fact, given the deep uncertainty inherent in climate projections and our growing understanding of nonlinear feedback loops (commonly referred to as "tipping points"), there are good reasons to proceed cautiously and to retain the catastrophe insurance framework as a complementary heuristic. Even when central tendencies of many climate models converge to lower levels of predicted warming than before, the tails of the probability distributions remain uncomfortably long and subject to deep (also known as Knightian) uncertainty, which could be summarized as a problem of unknown unknowns.

But there are serious psychological challenges of motivating sustained engagement based on doomsday scenarios. Moreover, they offer little actionable intel as to the practical challenges of adapting to the warming that is inevitable. It therefore seems imperative that we expand the set of mental models we employ as decision-makers and voters on the issue of climate change. That, at minimum, we consider the narrative heuristic of climate change as a slow and unequal burn—not so much as an alternative, but a complementary one to that of climate change as existential crisis for life on earth.

Of course, economic approaches to environmental problems risk the tyranny of measurement: we are prone to look for our keys under the lamp post because that's where the light shines. What we cannot measure well or quantify in dollars and cents may not get valued as highly, despite immeasurable intrinsic value. This risk notwithstanding, I hope to show that even when limiting our analysis to things we can measure reasonably well, the benefits of many clean energy investments very likely far exceed the costs.

As we will see in the pages ahead, the data now suggests that reducing emissions, even at reasonable cost, is probably worthwhile due to the preponderance of what I am calling the slow burn. This may be the case even before one considers existential tipping points in the climate system, or the very hard-to-value destruction of the natural world, such as biodiversity loss or the degradation of sacred groves. The lower probability extreme outcomes may well be increasingly likely and potentially devastating, and the as-yet poorly measured value of natural capital possibly very high. But I hope to make the case that, even when focusing squarely on the direct human losses associated with noncatastrophic warming in the near term, the cost-benefit of climate mitigation may favor aggressive emissions cuts.

A second flawed assumption in most climate change discussions is that the damages from climate change are somehow globally applicable, like an extinction-level asteroid or an alien invasion. That there is one universal phenomenon that is climate change, a riptide that affects all of us on planet earth equally. Here humanity sits, precariously balanced on a cliff, a modern-day Noah's ark. "We will all act together, or we will all die together" is often the implicit motif.

While such imagery may provide useful narrative devices, it is inconsistent with the available evidence. Especially when one considers how the rich and poor differ in their exposure to (and ability to adapt to) climate risk. Indeed, the previously listed facts one (immediacy) and two (inequality) necessitate a way of thinking about climate change that is less about all-or-nothing planetary poker, and more about the actions that may need to be taken to manage the warming that is inevitable during our lifetimes, and perhaps helping to shield the most vulnerable from its blows.

Both the perspective of the ark and that of pain management can be entertained—and the respective goals of mitigation and adaptation pursued—in parallel. One can, out of concern about the risk of fatal kidney failure in the future, decide to make swift and persistent changes to one's diet and exercise routine while simultaneously seeking immediate medical attention to address the progressing symptoms of chronic kidney disease now.

As I hope to illustrate through a series of studies and examples, these long- and short-term approaches—which together amount to walking and chewing gum at the same time—may become especially important when one considers just how unequal the realized effects of climate change already are.

This is not only because some parts of the world will warm faster than others or be more prone to flooding or hurricanes or wildfire. It is also because the way a given climatic event translates into realized human suffering (or flourishing) is almost always a highly complex function of the social, economic, and institutional environment in which it plays out.

A collision at thirty miles per hour in a Mercedes with seat belts fastened may lead to a few bruises and some insurance paperwork. Colliding at the same velocity in a rickshaw without a helmet may prove fatal, especially if there are no ambulances or hospitals nearby. So it is with many climate shocks as well. It turns out that the mortality consequences of a 95°F day can vary by a factor of ten or more between a place like India and a place like the United States. Even within the United States, the effect of such heat on health may vary by an order of magnitude, depending on highly localized and individual factors such as income, occupation, industry, and neighborhood.

Similarly, affluent homeowners may be able to erect protective bulwarks against storm surges and wildfires, or finance the move to safer terrain, while the poor may become stuck in increasingly natural disaster-prone places with eroding balance sheets that make them even less able to move out of harm's way.

However, as we peel back the layers of the onion on climate vulnerability it becomes clear that physical investments such as seawalls, air-conditioning, and more tree cover are only a small part of the solution set. In fact, in many of the world's poorest places, the most important adaptation interventions may ostensibly have little to do with the environment. We will explore the perhaps unexpected influences of banking systems, education systems, and labor markets in absorbing or amplifying the adverse effects of climate change on our well-being.

An explosion of research, much of it coming online over the past few decades, is quickly recasting our understanding of the hidden but pervasive ways in which temperature and extreme weather already affect our daily lives—thanks in part to parallel developments in the availability of data and the so-called credibility revolution in economics.

This book is largely a product of engaging with such data and the new quasi-experimental methods that help unlock insights hidden in them: both in my own research and as a member of a growing cadre of applied economists working on what one might call the microeconomics of climate change. This more data-driven, statistically minded manner of climate discourse will be essential to making more informed decisions about climate change.

Between 2010 and 2021 alone, there were over 400 peer-reviewed studies that combine real-world data and quasi-experimental research designs to shed light on climate damages.[13] These are studies that use techniques that enable us to better disentangle cause and effect (e.g., does hotter temperature actually lead to increased crime locally?) and that use data to inform our understanding of how societies respond and adapt or not. Among other things, these improvements help us to better identify the specific populations and places most vulnerable to climate shocks, as well as the potential effectiveness of various interventions. Much of what is practically valuable from these analyses may not yet be fully incorporated into policies or individual decision-making, despite potentially significant practical value.

There are no doubt many pitfalls in relying on past data to inform the future, particularly for a phenomenon as unprecedented in scale and speed as climate change. I hope to be mindful of such limitations early and often, and by no means does this book attempt to engage in anything resembling economic forecasting. These pitfalls notwithstanding, this book is written as an invitation to consider an economically nuanced perspective on climate change as, at the very least, a useful complement to other ways of perceiving and acting on the problem.

The Slow Burn and Its Silver Lining

Most professional athletes, musicians, and craftsmen will tell you that their prowess is a product of years of dedicated training, even though they may be naturally gifted in some way. Successful businesswomen and men know that, alongside their flashes of entrepreneurial insight were hundreds of incremental innovations that built the company's success over time. Spiritual teachers in Buddhism and elsewhere often refer to the process of "sudden awakening, gradual cultivation."

The obverse can also be true. Doctors know that, for every freak accident that lands a patient on their operating table, many arrive as the cumulative consequence of a million small choices: the extra cigarettes smoked; the midnight oil burned; the side of fries chosen instead of a salad. Bankruptcy attorneys will tell you that, while some personal bankruptcies are instigated by the financial strain of a sudden uninsured disaster, many are the cumulative consequence of daily habits.

Of course, there isn't quite a mirror symmetry here. Whether for bodies, businesses, or credit, it's usually easier to destroy than to build.

But a possibility this book invites us to consider is that, when it comes to climate change, much of the real damage may come not from the spectacular disasters but from the quiet, slow burn. The less salient, incremental disruptions spread out across a million seemingly mundane activities, processes, and interactions. That the more appropriate allegory to hold in our minds is not so much Mount Vesuvius and the annihilation of Pompeii, but the melange of epidemics, skirmishes, and managerial missteps that eventually crippled the mighty Roman Empire.

If the story of Pompeii is one of instant and catastrophic annihilation, the story of the fall of Rome is one of gradual decline. The former, while ostensibly more relevant to climate as a story of environmental destruction, has little to teach us about climate change. The latter, while a messier and more ambiguous affair, may turn out to be the more appropriate mental model as we confront what climate change means for our social and economic reality.

There is an important silver lining that emerges from this perspective. It is the recognition that there are more ways one can make a positive difference than first meets the eye, particularly in helping the most vulnerable communities adapt to a warmer future, whether they reside in the world's poorest rural hinterlands or the most dilapidated neighborhoods of your hometown.

Whereas the binary framing of climate change as impending catastrophe incites fear and fatalism, the subtleties of noncatastrophic climate damages just might, I submit, instill compassionate resolve and perhaps even informed optimism. It can show us how we are never too late to slow the progression and to better manage the pain of a warming planet. It might just help steel our collective determination for the decades of hard work ahead—that is, for the work of transitioning our societies and economies toward cleaner and more climate-resilient versions of themselves.

My own hope is that this book will spark greater curiosity regarding the possible interventions, individual and societal, that could help manage the complex interplay of climate change and the inequities and uncertainties of modern economic existence. Additionally, I hope it might help us to be more curious about why a hotter climate can be more damaging to some than others and to learn how we might invest in the knowledge, policies, and processes that will help blunt climate change's inevitable blows. We all may be surprised to learn how poorly adapted some communities are to the current climate (let alone the future one), and how the most cost-effective adaptation strategies may have seemingly nothing to do with climate.

It is easy to demonize others about the dire straits we find ourselves in: The oil companies who chose to obfuscate rather than educate; the politicians who put expediency above principle; the earlier generations who grew fat and complacent on the sugar high of a fossil fuel–intensive economy.

The truth, however, is more uncomfortable. The truth is that we are all complicit in the fossil-fuel economy to some degree. Moreover, the way our minds incline hinder us from grasping the scale and abstraction of the problem in an actionable way. The fact is that as humans we are not naturally inclined to think statistically or probabilistically about climate change (or for that matter, most policy problems).

As such, we may all be predisposed to overlook crucial pieces of the puzzle, including the understanding that, despite it being ostensibly a physical environmental phenomenon, much of what determines whether climate change hurts or helps has to do with the complex interplay between environmental and economic systems. The new data and perspective offered in this book underscore in particular the mitigating or exacerbating influence of economic context. What hurts us are not simply the natural phenomena of hotter temperature, rising sea-levels, or more variable rainfall, but how they interact with human institutions—economic, educational, legal, and political.

I want to be careful not to make this yet another *moral failing*. If anything, part of the difficulty we face may be the morally charged tenor of much climate change discussion. What I am referring to here is less a moral failing, more an issue of our *mental foibles* and how they influence the way we think and talk about climate change.

Caveats of a Two-Handed Economist

A frustrated US president Harry Truman once demanded that his aides find him a "one-handed economist," lamenting the tendency of his economic advisers to caveat everything with "on the one hand . . . on the other."

Lest you think this economist was going to buck that trend, here are some important caveats before we proceed.

In the chapters that follow, I appeal to vignettes and anecdotes frequently. This book is intentionally long on intuition and short on equations. This is mainly to make the ideas more digestible, to help the concepts stick.

While anecdotes are powerful tools for communication, they are not a substitute for rigorous scientific evidence. Anecdotes can be chosen to make a particular causal story more intuitive, but it is not the story that makes the causal statement true. We rely on the scientific method to provide us with generalizable facts about the path between cause and effect. That, for instance, a particular vaccine causally reduces the likelihood of hospitalization and death. That, given how the assessment was conducted (e.g., a randomized controlled trial of sufficient size and duration), the arrow of causality runs in a particular direction—vaccine causing improved immunity—and is probably not being driven by other unobserved factors or by selection bias (e.g., healthier individuals having been more likely to take the vaccine).

It is therefore with great caution and care that I offer the anecdotes and vignettes included in the pages that follow. Unless otherwise stated, or clearly offered as allegory, they are chosen with the understanding that the narratives presented are consistent with the results of peer-reviewed scientific research. Of course, there will be limited space to discuss the details of how we know what we know and with what degree of certainty. For the reader who would like greater technical substantiation, I have included endnotes where I reference and sometimes expand on the studies that form the basis of such assertions and at times describe their data, methods, and specific strengths and weaknesses in greater detail, as well as mention other studies that allow you to delve even deeper should you find something to be of particular interest.

Along similar lines, both the coverage of this book and the applicability of its central idea are inherently flawed and incomplete. My intention is to introduce a different way of looking at the issue of climate change, not to provide a comprehensive overview of the subject. As such, there will invariably be major aspects of climate change—or parts of the world affected, different disciplinary perspectives to be considered—that are discussed only sparingly, or not at all. And as with all heuristics, the ones I offer here are imperfect at best; at worst, hindsight and more research may reveal them to be grossly misguided.

Finally, the book also does not pretend to provide definitive answers to all the questions raised. In some cases, what the data reveals will be unsatisfyingly ambiguous, simply giving rise to different questions. Hopefully I have managed to be clear about where there is general consensus within my field, and where I am merely providing food for thought. For readers who would like a more methodical treatment of the levels of consensus and uncertainty, I would recommend the latest Intergovernmental Panel on Climate Change reports, as well as country-specific analogs such as the National Climate Assessment in the United States or reports by the UK Climate Change Committee.

PART I

1

Thinking Fast and Slow about Climate Change

NOBEL PRIZE–WINNING economist Amartya Sen has argued that close to 100 million girls and women appear to be "missing" globally. This implies that more girls may have been killed in the past fifty years simply because they were girls than men were killed in all of the battles of the twentieth century, as suggested by Nick Kristof and Sheryl Wu-Dunn in their best-selling book *Half the Sky*.[1]

Let us reflect on how many people that is. All the German, French, British, and Russian soldiers perishing in the mud during the trench warfare of World War I; all the male naval officers in Pearl Harbor; and all the men living in Hiroshima and Nagasaki. The numbers are almost too large to wrap one's mind around: there were an estimated 21 to 25 million military deaths in World War II alone.

The broader point is that, because of the subtle disadvantages of being born a girl, millions of human beings have been dying largely silent deaths all around the world, whether through infanticides, sex-trafficking, or the cumulative disadvantages brought about by educational disinvestment.[2]

Why then weren't more people aware of this phenomenon? Were you? Kristof and WuDunn describe the tendency this way:

When a prominent dissident was arrested in China, we would write a front page article; when 100,000 girls were routinely kidnapped and trafficked into brothels, we didn't even consider it news. Partly that is because we journalists tend to be good at covering events that happened on a particular day, but we slip at covering events that happened every day—such as the quotidian cruelties inflicted on women and girls. Journalists weren't the only ones who dropped the ball on the subject: a tiny portion of US foreign aid is specifically targeted to women and girls.[3]

Could something similar be at play with climate change? Could it be that the way climate change tends to be covered in the press and discussed at the dinner table hinders us from grasping the true extent of its human consequences?

According to the Yale Project on Climate Opinion, 72 percent of Americans believed as of 2020 that "climate change is happening." Sixty-six percent of adults aged 18 and older said they were worried about climate change. The vast majority (73 percent) also said that global warming "will harm plants and animals."[4]

These are significant shifts from even a decade ago. Even in historically climate skeptical Australia, more than 60 percent of respondents believed as of 2020 that climate change is a serious issue, compared to 36 percent in 2012.[5]

According to the same Yale survey, only 43 percent of Americans believe that climate change will harm them personally. In much of rural America, that number appears closer to 30 percent.[6]

While the vast majority now agree that climate change is happening and is a problem, most in the developed world still believe that it is a problem mainly for other people, other sentient beings, or something more nebulous like "ecosystems," "the planet," or "Mother Earth." In many of our minds, most of the time, even the people who will be harmed reside far away, somewhere in far flung rural reaches or small-island nations.

Just as importantly, it appears that many, even those inclined to view climate change as a serious problem, may not be predisposed to appreciate the aspects of it that are most harmful. This, as we will discuss, may have significant implications for how we make a range of important decisions, societal and personal.

This might also have something to do with the fact that climate change seldom rises to the top of the priority list. According to a 2022 *New York Times*/Siena College poll, 1 percent of American voters named climate change as the most important issue facing the country, well behind worries about inflation and the economy. Even among voters under 30, the group thought to be most energized by the issue, that figure was 3 percent.[7]

Thinking Fast and Slow

There is a kind of cognitive dissonance that comes with being a scientifically literate person in the twenty-first century, one that does not get as much attention as it deserves. It is the tension between the more instinctive responses of our neurological hardwiring—the ability to read emotions on the faces of others, or to jump out of the way of an oncoming truck, for instance—and the more deliberative thought processes of the rational mind.

Perhaps it is a version of what psychologist Daniel Kahneman calls "thinking fast and thinking slow." In the book *Thinking, Fast and Slow*, Kahneman points out some of the systematic blind spots of human intuition, summarizing a vast collection of research led by himself, Amos Tversky, and others.

To crudely paraphrase this careful work, the mind can be thought of as having two main players, or systems: the more intuitive, "fast thinking" mind and the more deliberative, "slow thinking" or so-called rational one. As Kahneman and Tversky call them, the fast thinking mind is "system one" and the slow thinking mind is "system two."

One of the reasons why wrapping our heads around climate change is so difficult may be because it forces us to think statistically, which is squarely in the domain of the slow thinking mind. As Kahneman puts it: "Why is it so difficult for us to think statistically? We easily think associatively, we think metaphorically, we think causally. But thinking statistically requires us to do many things at once, which system one is not designed to do."[8] Probing whether individually imperceptible changes in productivity associated with an outdated payroll software or subway line or education system materially affect a company's bottom line or a city's municipal budget or a nation's rate of economic growth over time requires, at minimum, thinking statistically. In the context of climate change, assessing whether the unusual weather experienced as of late is a chance occurrence or meaningfully connected to a changing climate requires thinking statistically. Discerning whether my own experience of diminished mental acuity on a hot day (or in an overly air-conditioned room) is a generalizable phenomenon or personal idiosyncrasy requires thinking statistically.

Pulling all the pieces together to assess the costs and benefits of climate action—that is, determining whether 1.5°C of warming differs from 3°C or 4°C or more, and how those differences compare to the societal costs of achieving the necessary emissions cuts over time—all of this requires us to think statistically. (It also likely requires a credible way of distinguishing cause and effect in the context of multiple, potentially correlated causal factors, as well as a means of quantifying in dollar terms many important aspects of human experience that are not easily monetized, among other things.)

Of course, human intuition is an incredibly powerful tool, particularly when placed in the right context, and armed with the right inputs and training. The more intuitive system one is usually very adept at drawing associations between familiar objects, or judging what is or is not a fair outcome. It can also become highly skilled, once an initial training period has passed, at effortlessly engaging in what are actually quite complex tasks like tying shoelaces or diagnosing a tumor or composing a sonata.[9]

But intuition is typically only trustworthy when people build up experience making judgments in a predictable environment.

The point I am trying to make is that certain domains are simply not well-suited for system one's default cognitive processes, and that many aspects of climate change fall under this category.

Behavioral scientific research indicates that there are many situations in which relying on system one's intuitive judgments tends to lead us systematically astray. For instance, consider the availability heuristic. This refers to the mental shortcut by which we rely on immediate examples that come to mind, even when they might not be the most appropriate or relevant in evaluating an issue or making a decision. News of high-profile subway shootings may make more people drive to work, even though driving still constitutes a more dangerous mode of transport statistically. Most people, when asked which is a more dangerous occupation, logging or policing, will answer the latter based on the availability of mental images associated with cops in the line of fire, even though logging results in more serious injuries on average.[10] There are almost too many cop films to count. How many Hollywood blockbusters are about industrial loggers?

Because our expectations about the frequency of events are distorted by the prevalence and the emotional intensity of the messages we receive, it seems possible that we may be simultaneously slow to consider climate change a serious problem and quick to focus on misleading aspects of it when we do.

Of course, trying to circumvent or double check our intuitive judgments and processes with constant vigilance is exhausting, and probably impractical. Here, Kahneman and Tversky suggest an important compromise: Try harder to avoid significant mistakes when the stakes are high.[11]

A starting supposition of this book is that, when it comes to decisions about climate change—especially collective decisions around public policy—the stakes are very high indeed. Climate change is a problem that involves thinking statistically on many dimensions, and arguably bedevils our more intuitive processes in many ways, yet its planetary scale and the limited time available to reverse course instills today's decisions with potentially immense consequences. A goal of this book is to show that the stakes may be higher than commonly assumed—both at a societal level and an individual level.

Cognitive Biases and Climate Change

There are many reasons why climate change poses a unique social challenge. Most are at this point well-chronicled and may be familiar to the reader.

Economically, it represents a colossal market failure. As the pioneering environmental economist Martin Weitzman once put it: climate change constitutes the "mother of all externality problems."[12]

In case it has been a while since you picked up a microeconomics textbook (or you never had to), a negative externality arises whenever some transaction, whether an investment or a purchase or a service rendered, imposes costs on a third party (or third parties) not privy to that transaction. The sticker price for gasoline at the pump does not fully reflect the costs borne by the rest of society: that is, in the form of local air pollution, traffic congestion, and global climate change. So we end up driving more and taking mass transit less than would be socially ideal. Of course, externalities aren't unique to the environmental realm, nor are they always negative. The noise of a frat party or a constantly barking dog imposes negative externalities on the neighbors. The historic chestnut tree in your backyard or the lady who picks up stray trash on her evening walks generates positive ones. But the climate externality is likely of a scale and magnitude hitherto unprecedented. Across billions of everyday decisions, carbon-intensive options are inadvertently favored over ones that are less damaging to the atmosphere.

The externality problem of climate change is further exacerbated by time: the fact that climate change unfolds globally over many generations. This means that, in addition to there being hidden costs associated with my actions borne by billions of other people *today*, there are arguably larger costs for billions of other people *tomorrow* and many years from now. In fact, to the extent that the climate change generated by my emissions today will affect future generations who do not have a say in whether I drive a Hummer or a hybrid or take a bicycle to work (and given that even the most altruistic among us may not fully incorporate the needs of all of posterity in every decision we make), the externality effects noted above are amplified due to this intergenerational ripple effect.

Politically, it is a major collective-action problem requiring global coordination on a similarly mammoth scale. The fact that we all share the same planetary climate means that the costs of cleanup (i.e., reducing the accumulated global concentration of greenhouse gases) make an environment ripe for free-riding. One could crassly paraphrase a rich political economic literature with the following aphorism: Why should I pay the costs for cleanup when you continue to trash the place? In some ways, it is the perennial challenge of keeping a common area of a college dorm clean, but to the nth degree—across all the world's cultural, economic, and linguistic differences.

In other words, climate change suffers from what Nobel laureate William Nordhaus calls the "nationalist dilemma" and the free-riding it begets: countries that do not participate in a global agreement to reduce emissions (or chip in less than others do) get a free ride on the costly abatement by other countries.

Even when there is an interest in understanding and doing something about the climate problem, there are additional psychological challenges to overcome. To start, the most severe impacts from climate change—sea level rise of several meters, for instance—will probably occur in the very distant future. Most of the sea-level projections that include the melting of the Greenland ice sheet are based on a steady state that is at least 200 years from today.[13] This makes it inherently difficult to fathom the gravity of the situation. Something else always seems more urgent, because, in an individualistic, myopic sense, it usually is.[14]

All of these reasons have been discussed in detail by minds more astute than my own. However, my particular professional keyhole into the climate problem suggests that there is another challenge, one that may be similarly obstructive of change.

It is perhaps a subtler psychological handicap that arises from our tendency to gravitate toward headline-grabbing catastrophic imagery while being more sanguine about the routine events, combined with the difficulty of credibly assigning cause and effect to experiences that occur with varying frequency. There are many additional elements at play in this phenomenon, including the way the post–social media landscape incentivizes the memeing of more extreme and simplistic positions. But the closest thing to an elegant shorthand is Kahneman and Tversky's problem of thinking statistically.

For instance, without the temperance of statistical thinking, our senses are bedeviled by natural variation. Weather is what we experience day to day. Climate is an average tendency of weather over a longer period of time. Weather naturally fluctuates, sometimes in directions that ostensibly contradict the slow upward march of global warming.[15] For instance, there may in some places paradoxically be more frequent days that feel unusually cold because, as the world warms, the jet stream can weaken, leading to periodic collapses that permit arctic air to travel farther south than usual. This is already leading to extreme cold events, like a spell of subzero days in late springtime for many parts of the temperate world. It is also creating confusion, and opportunistic political showmanship that feeds on it.[16]

Once again, this is itself an admittedly compromised heuristic. There is no single term that does justice to the complex array of phenomena I am describing here. I use the term *thinking statistically* to encapsulate a number of other related ideas as well, like thinking probabilistically and in terms of population averages as opposed to individual experiences. Also required to think clearly about these issues is a high tolerance for abstract reasoning, and an ability to think about longer time horizons than typically required in everyday life. And, as we will see in the pages that follow, even while thinking statistically, there are many factors that complicate attempts at an accurate understanding of climate risks. For instance, when it comes to understanding the real costs of climate change, our senses—and our societal information-gathering systems—may be systematically biased against registering some of its most consequential facets.

My claim in this book is that this bias may have very real consequences for how we understand the urgency of the climate problem, and possibly whether and how we act on it as well. Could it be that many of the costs of climate change are hidden from view and thus evade news coverage, or even traditional recordkeeping? When climate disasters hit, the physical destruction of roads and bridges makes for striking headlines. But what do we know about the hidden effects on learning and human capital? When wildfires engulf California or Catalonia or Australia, should we actually be just as worried about the smoke as the flames?

Unpacking the stories behind these clandestine costs, their potential aggregate magnitude, and how we as a scientific community have attempted to overcome our collective blindness are major objectives of this book.

But let us focus for now on how this bias plays out in our basic conception of the climate problem—of how bad the consequences are and for whom, and how that informs our decisions around climate action. In the next two chapters, we will explore this idea further by focusing on what headlines may miss about natural disasters such as hurricanes, floods, and wildfires.

2

Physical versus Human Capital

Superstorm Sandy

In October 2012, Superstorm Sandy hit the northeastern United States, flooding streets and homes, cutting power to millions, and causing extensive property damage from Florida to Maine. Some of the most enduring images from the storm's aftermath come from New York City, where widespread power outages combined with flooding of subway tunnels to create a scene out of some dystopian horror movie. As a former New York City resident, it is difficult to fully scrub from my mind footage of entire blocks of Lower Manhattan underwater, or of sewage-infused storm water overflowing subway platforms.

Of course, such damages were only the more iconic tip of the iceberg: downed power lines meant over 6 million residents went without electricity across Washington, DC, New Jersey, Pennsylvania, and Connecticut; major transit systems and airports were shut down for days; thousands of homes and businesses suffered property damage. All told, official estimates put the economic damages from Hurricane Sandy at over $85 billion, making it one of the costliest hurricanes in history.[1]

What is less often appreciated is how disruptive Hurricane Sandy was for young people. The storm shut down all 1,750 New York City public schools, halting learning for over 1 million students for an entire week, likely causing pedagogical ripple effects that lasted far longer.[2] Schools in Washington, DC, Boston, Philadelphia, and Providence, Rhode Island, were temporarily shuttered as well.[3]

In New York, power was not fully restored in some areas for weeks. Many New York City students were relocated to temporary facilities while their classrooms were being repaired. In some cases, students were not able to return to their normal school locations until the following year. Because of the disruptions to subways and bus networks, many parents had difficulty getting their children to the new school locations. Even in the schools that reopened, attendance fell by as much as 20 percent in the weeks following reopening, relative to prestorm averages.[4]

The $85 billion figure cited above does not account for any such learning disruptions.[5] This is understandable. There is no straightforward way to account for such costs. Plus, the narrative mind has a hard time playing out the counterfactual. How can we really know whether a few weeks of disrupted classes are really a big deal in the grand scheme of a child's education?

The Human Capital Consequences of Natural Disasters

When environmental disasters hit, the damage to physical capital—so-called brick and mortar assets—is usually quite apparent. Downed power lines; bridges washed away by floods; the charred remains of homes and cars in the wake of wildfires. We see such footage in the news all the time.

Less obvious is the harm done to human capital, which, broadly speaking, can be thought of as the capacity of human agents to engage in productive economic activity. In an increasingly knowledge-based economy, the skills and capacities learned in school comprise an asset base every bit as essential as the bricks in a building or the steel rods in a factory.

Could it be possible for largely invisible disruptions to human capital formation to have economic implications that are deeper and more impactful than first meets the eye? This is important because educational attainment and the skills that schooling and training confer have become an increasingly valuable input to success in the modern economy. This is true both in terms of aggregate economic performance—the relative wealth and poverty of nations—as well as for individual economic mobility, that is, for the chances of moving up the economic ladder within a given society. As we will discuss in greater detail in later chapters, the implications of climate change for economic inequality may hinge in part on how readily individuals can access the education and skills necessary to compete in an increasingly technologically complex and automated workplace.

Physical capital assets tend to be much easier to value monetarily, in part because they usually comprise tangible quantities traded on the market. You could probably look up the estimated trade-in value of your car in a matter of minutes on your smart phone. Even in the case of bridges or airports, which are not bought and sold on markets like homes and cars, the cost of repair—in terms of materials and construction labor—can be readily calculated.

Human capital, on the other hand, is often less tangible, and certainly far less fungible. Indeed, there are a great many such intangible assets that are vital to human flourishing. These include not only educational achievement and job-specific skills, but also mental health and physical vitality (what economists call *health capital*). One can also think of things like the air and water filtration benefits of nature (known as *ecosystem services* or *natural capital*), and even the historical significance of a particular building or monument or the degree of social cohesion or sense of safety one feels in a neighborhood, as forms of intangible capital that, while not actively traded on markets, are foundational to our quality of life.

What do we know about the effects of climate on such intangible, nonmarket assets? Increasingly, the evidence suggests that headline damage estimates—which typically focus on physical capital—may miss a wide swath of economic harm. In chapter 7, for instance, we will focus on what the new evidence tells us about how climate change affects the tenor of our social interactions, and whether a hotter world may be a more irritable, error-prone, and conflictual one.

Let us focus for now on one component of intangible human capital: formal education.

Costing Katrina

If a storm knocks all the apples out of a fully grown apple tree before they are ready to be picked, the losses are significant, but with some care the harvest can bounce back the next season. If a storm damages the roots of a growing seedling, however, it may eventually mature to produce fewer or more often diseased apples every year than otherwise might have been the case. Over the lifetime of an apple tree, the difference in the cumulative number of harvested apples may turn out to be very different between the damaged seedling and the average tree. Could there be something similar about the way children learn and grow that makes them especially susceptible to adverse shocks—natural disasters included?

Hurricane Sandy represents only one of over three hundred weather and climate disasters that the National Centers for Environmental Information reports as having caused at least $1 billion in damages since 1980. Consider a different, even more damaging storm: Hurricane Katrina, which is reported to have caused $175 billion in damages, making it by far the costliest hurricane on record.

Careful analyses of individual records for tens of thousands of Louisiana public school students reveals that test scores of students who evacuated New Orleans due to Katrina fell by anywhere between 7 percent and 20 percent of a standard deviation that following spring, which is in the same order of magnitude as the positive effects of attending a charter school.[6]

This is not a small effect, and perhaps unsurprising given that students in New Orleans' public schools ended up missing five weeks of school due to the storm. It is difficult to imagine students being able to focus on learning when they and their families are forced out of their homes due to flooding, with seventeen out of the nineteen New Orleans zip codes prohibited from returning to their homes until nearly four months after the storm.

To put the learning impacts in perspective, these effects are on par with what would happen if we replaced average teachers with those one to two standard deviations below average—that is, teachers in the bottom 10 percent of the teaching ability distribution—for an entire school year. Here's another way to think of it. Across a wide range of studies, we find that increasing per pupil spending by $1,000 for a sustained period (four years or more) increases learning by 3 percent of a standard deviation per year.[7] So, if the effects are symmetric, Katrina's effect is on par with reducing per pupil spending by at least $2,000 each year, possibly much more.

A perhaps unexpected silver lining in the case of Katrina is that students who relocated to districts outside of New Orleans ended up attending much better schools in better neighborhoods. In part because of the scale of devastation and the media and political firestorm that followed, Katrina generated an unprecedented funneling of national attention and resources that helped relocate a far greater number of people than typically occurs after a storm.

This appears to have, over the course of several years, more than made up for the lost learning due to the initial disruption. (Indeed, even the elderly who were relocated appear to have benefited in the long run from improved health, possibly due to better healthcare systems in the cities that they relocated to.[8]) This may be small consolation for these students and their families who had to uproot and leave their home communities. But as we will see in later chapters, such natural disaster–induced relocations may be closer to the exception than the norm. It seems possible that, in many cases, people go back to where they were before the storm, just with fewer resources to work with, more rebuilding to engage in, and more makeup classes to attend.[9]

Two facts are worth emphasizing here. The first is that, due in part to the media attention it generated, Katrina garnered a mobilization of external support unique in its scale and magnitude. The second is that, in spite of this, the total price tag of $175 billion in economic damages often quoted in the press does not include any of the later-life earnings implications of such acute disruptions to student learning—which, for better or for worse, is one of the easiest ways to measure loss of human capital.[10]

Quantifying Hidden Human Capital Costs

Katrina and Sandy are two very extreme events. In fact, they rank first and fifth, respectively, on the list of all-time costliest storms in the United States. What of the many hundreds of far less extreme events?

Economists Isaac Opper, Lucas Husted, and I were interested in what happens in the case of less obviously catastrophic natural disasters. We were particularly interested in how the human capital consequences of such run-of-the-mill natural disasters might stack up in dollar terms.[11]

Fortunately for us, the US government takes great pains to document the property damage associated with all manner of natural disasters, ranging from hurricanes and floods to tornadoes and wildfires. Because federal aid dollars are administered on the basis of assessed physical property damages, government agencies spend considerable time and resources trying to tally the dollar value of damaged and lost property for any given natural disaster. Figure 2.1 provides a map of all presidential disaster declarations, shaded by the frequency and amount of assessed physical damages in dollars per capita (small: $0–$10; medium: $10–$100; large: $100–$500; very large: $500+).

Most natural disasters that are officially given presidential disaster declarations impose property damages of less than $500 per capita. Just to give you a sense of how much of an outlier Katrina was, the federal government estimates it to have caused approximately $34,000 in damages per capita. The Oklahoma tornadoes of 2021 caused $6 to $111 in damages per capita among the five affected counties.

Unfortunately for us, there is far less information on the human capital damages. So we pulled together data on standardized student achievement for nearly all third through eighth graders in the United States (available thanks to many years of data collection and harmonization on part of researchers at Stanford), as well as information on high school graduation and post-secondary enrollment rates across the country between 2008 and 2018. We combined this with information from the universe of presidential disaster declarations over a roughly thirty-year period, as well as detailed local property damage assessments.

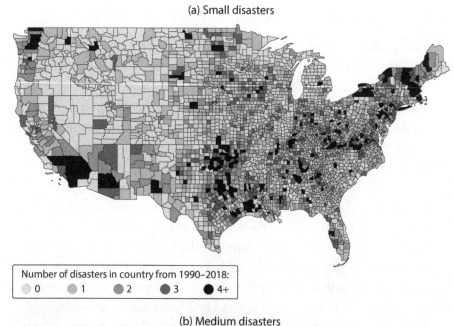

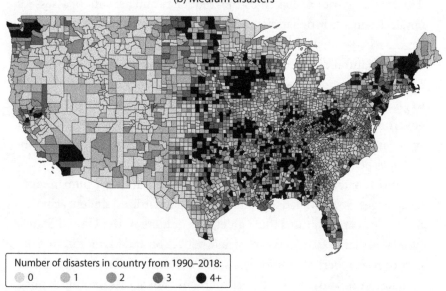

FIGURE 2.1. These maps show the number of times a county was impacted by disasters in the period from 1990 to 2018, using data from federal disaster declarations: (a) small disasters, which consist of counties with per capita property damage of $1–$10 per capita in a year; (b) medium disasters, which have per capita property damage of $10–$100; (c) large disasters, which have

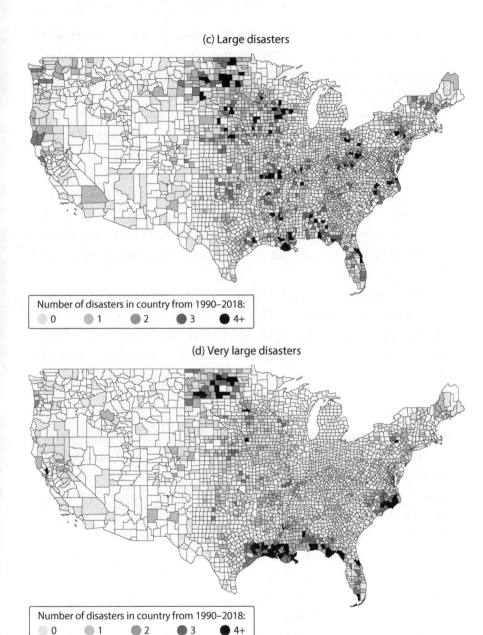

FIGURE 2.1. (*continued*) per capita property damage of $100–$500; and (d) very large disasters, which have more than $500 per capita property damage. *Source*: Isaac M. Opper, R. Jisung Park, and Lucas Husted, "The Effect of Natural Disasters on Human Capital in the United States," *Nature Human Behaviour* (June 2023): 1–12, https://doi.org/10.1038/s41562-023-01610-z.

We measured the impact of natural disasters of varying magnitudes on these two measures of human capital: student achievement, in the form of gains in test scores, and educational attainment, in the form of changes in high school graduation and college enrollment. We exploited the fact that many places in the United States are hit by multiple types of natural hazards of varying intensity, repeatedly over time. For example, we could compare the gains in learning that occur in a particular school district in the Chicago area in an average school year to the gains that occurred in that same Chicago district in school years where students were exposed to a natural disaster, also controlling for factors that may have affected learning across the board (like the Great Recession), as well as potential trends in disaster frequency and local economic and demographic trends.

Our hypothesis was that, by being forced to cope with a potentially wide range of additional disruptions to learning, students who experienced natural disasters during the school year may eventually perform worse on exams aimed at testing their knowledge of the subjects studied during that time. We also wondered whether natural disasters made it less likely for students to complete high school or to attend and stay in college.

It is worth reminding ourselves what the process of learning entails. Learning new concepts and skills can be difficult enough in the best of circumstances. It requires time, effort, and a reasonably distraction-free environment. Sometimes, in order to be able to learn the next concept, you need to have mastered the first. So disruptions at one point in time may lead to further compounding learning difficulties in the future.

Consistent with other work, we find that, across a range of disaster types, natural hazards can indeed be highly disruptive to learning. Test scores drop significantly in school years where students experienced a natural disaster. College enrollment drops as well. In both cases, disasters in future years have no impact on current test scores or college enrollment, which is what we would expect if year-to-year variation in disasters was in fact more or less randomly assigned. As figure 2.2 shows, these educational impacts appear to scale in line with the assessed physical destructiveness of the event.

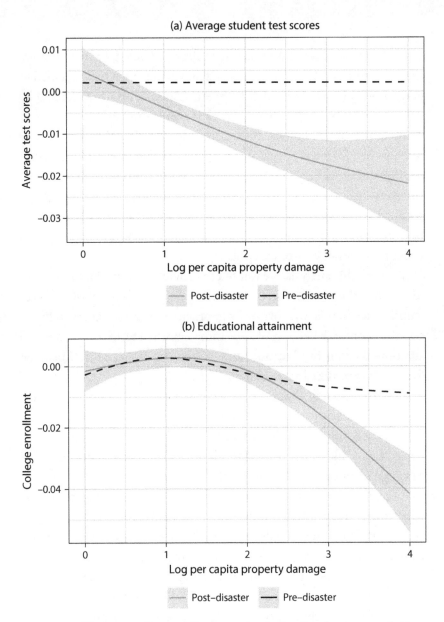

FIGURE 2.2. The impact of natural disasters on local educational outcomes, plotting the relationship between (a) average student test scores (standard deviations) and (b) college enrollment (log enrollment rates). The solid lines represent the effect of disasters of varying intensity (cubic splines), as measured by log per capita property damage, on average test scores and college enrollment rates within the affected county, with gray shading representing 95 percent confidence intervals. The dashed lines are results of a falsification exercise where prior year test scores are related to future disasters. *Source*: Isaac M. Opper, R. Jisung Park, and Lucas Husted, "The Effect of Natural Disasters on Human Capital in the United States," *Nature Human Behaviour* (June 2023): 1–12, https://doi.org/10.1038/s41562-023-01610-z.

High school graduation was not as clearly affected, for reasons that are as yet unclear. One possibility is that teachers may have decided to grade exams more leniently for upperclassmen who were affected by natural disasters. Such benevolent grade manipulation has been documented in a range of other studies, including in New York City, where teachers had for years been boosting scores of students just below passing thresholds to help them graduate on time. Another is that the disruptions we document are not large enough to meaningfully affect high school graduation on average, which in the United States is often subject to laws governing compulsory schooling up to a certain age, whereas college enrollment is not compulsory.

Through this analysis, we wanted to provide at least a ballpark sense of how bad the educational impacts are for a given amount of property damage. To do this, we borrow from previous studies that carefully assess the economic benefits of education—both in terms of educational attainment (years of education) and the content of that education in the form of standardized achievement.

One way to think about the monetary value of schooling is in terms of the gains made possible in terms of later life earnings. Setting aside the important question of to what extent the value of education extends beyond its value in the workplace, let us work with the premise that earnings advantage may be one useful metric of many.

As one can imagine, there are many reasons why simply correlating education and earnings across individuals is highly problematic, as higher educational attainment tends to be correlated with other things that may be beneficial in the job market, such as social connections or parental resources. To get around this problem, many careful academic studies have exploited natural experiments to estimate the causal effects of an additional year of schooling on earnings. These include quirks in compulsory schooling laws, random eligibility cutoffs, or twins being separated at birth, which lead otherwise very similar students to vary in the amount of formal schooling received, thus allowing researchers to hone in on the variation in later-life outcomes that might arguably be causally attributable to the additional education.

According to a recent review of many dozens of such studies, an extra year of schooling appears to confer an average earnings bump of approximately 10 percent per year.[12] So, for the average American with a median income of $44,000, that amounts to something on the order of an additional $4,400 each year over the course of one's career, or a net present value (NPV, which is the present-day equivalent of the expected future income generated by this "investment") of around $75,000. This reflects the gains associated with changes in final educational attainment, which often (though not always) becomes codified in the form of a terminal degree: high school diploma, associate's degree, bachelor's degree, master's degree, and so on.

Another way to try to monetize learning impacts is by looking at the effects of the actual learning itself: the value of learning important skills like critical thinking, writing, math, or science. Aside from the fact that treating learning as a financial investment potentially debases some of its intrinsic value, this is obviously also a very difficult forensic exercise, given the technical challenges not only of measuring learning and its retention, but also of linking learning gains to labor market outcomes that occur far in the future. A groundbreaking study by Raj Chetty, Nathan Hendren, and Jonah Rockoff uses administrative data from the Internal Revenue Service (IRS) and the New York City public school system to tackle this question.[13] Once again, exploiting quasi-experimental research designs and linking individual test scores to the corresponding students' tax records later in life (e.g., age 30), they show that students who learned more in a given school year appear to earn significantly more over a decade later, once the students have entered the workforce. For instance, having a one standard deviation better teacher improves learning by 13 percent of a standard deviation, which the authors show to result in approximately $7,000 in increased future earnings (NPV) per student.

We use both approaches, mapping the learning and educational attainment losses associated with a given natural disaster to potential later-life earnings losses. Because the data on property damages is so extensive, we can actually provide something that approaches an apples-to-apples comparison of dollar values: one set of dollar values for assessed physical property damages, another for the hidden human capital damages for the same set of natural disasters in a given county. These are imperfect for many reasons. But we think they provide a good place to start.

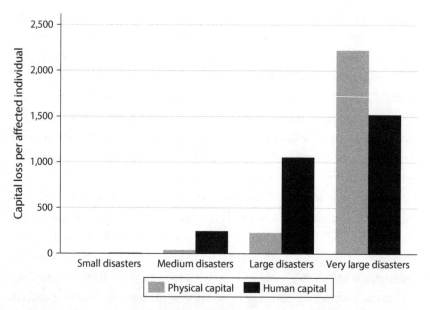

FIGURE 2.3. The estimated damage to physical capital and human capital of different-sized disasters. The human capital damages are computed as the cost per student; physical capital damages are computed as damages per capita. Small disasters consist of counties with per capita property damage of $1–$10 per capita in a year; medium disasters, $10–$100; large disasters, $100–$500; very large disasters, more than $500 per capita. *Source*: Isaac M. Opper, R. Jisung Park, and Lucas Husted, "The Effect of Natural Disasters on Human Capital in the United States," *Nature Human Behaviour* (June 2023): 1–12, https://doi.org/10.1038/s41562-023-01610-z.

As shown in figure 2.3, natural disasters that cause on average $10 to $100 per capita in physical property damage appear to result in human capital damages of approximately $41 per capita ($245 per student). The effects appear to scale more or less proportionately with the magnitude of disasters, at least when measured in terms of physical destructiveness. Disasters that cause more than $500 per person in assessed physical capital damages lead to human capital impacts on the order of $268 per person, or $1,520 per student.

This suggests that the learning losses may be in the same order of magnitude as the physical capital destruction.

Are these numbers exact? Certainly not. To be frank, we were a bit surprised by how large they were. We tried to think of a host of reasons why they might be too large. Some are worth noting here, like the fact that some students may move to better schools (though very few in our data appear to migrate in response to most natural disasters), or that learning gains driven by better teachers may also come with hidden soft skills and emotional intelligence that overstate the link between standardized achievement and earnings.

But in the end, we were convinced that they just might be in the same ballpark, and perhaps even conservative. For instance, our estimates do not account for the fact that the benefits of human capital may compound over time. Research increasingly shows that beneficial investments in earlier years such as kindergarten can confer learning benefits in subsequent years, leading to more rapid learning uptake akin to compound interest.[14] Conversely, negative shocks to younger children can be more destructive, as they often appear to have long-lasting ripple effects. We did not account for such dynamic complementarities, nor do we account for any effects of natural disasters on students younger than third grade, or potential direct health impacts on the students or their parents, which may have additional impacts over time. Given the growing evidence on how natural hazards and stress generally affect in-utero and infant health, and how such factors translate into educational achievement later during adolescence, our estimates might be understated.[15]

Our hope is that future researchers will help improve this somewhat crude first pass. How far off the mark this work proves to be is anyone's guess. But in some ways, these numbers seem plausible given the vital importance of education and the limited window of opportunity to make educational investments.

Hurricanes Sandy and Katrina were massive outliers in terms of their sheer physical destructiveness. For many residents, they were cataclysmic events. But these analyses suggest that the vast majority of less obviously catastrophic disasters may exact hidden costs that are not reflected in headline damage assessments, which typically rely on easier-to-measure property, infrastructure, and business-disruption costs. Zooming out, it seems possible that, when it comes to climate change, the less-readily monetized impacts on intangible capital—whether on student learning or health or something else like ecosystem services—may be of a magnitude warranting legitimate concern.

As we will see later in chapter 7, such intangible climate damages may be important for reasons beyond simple accounting. As the adage goes, what gets measured gets managed. Especially when one thinks about how best to insulate societies from the coming climate shocks to the system, having a more accurate sense of where damage is being done could be important in allocating and directing resources to help rebuild and improve resiliency. It is worth noting, for instance, that in the wake of hurricanes, inflows of government aid increase generally, but funding for educational assistance actually declines, for reasons that are as yet unclear.[16]

Heat and Learning

New evidence suggests that the hidden harm to human capital may also contribute to racial and economic achievement gaps, both in the United States and internationally. In a study of over 21 million PSAT exams from all across the United States, my colleagues Josh Goodman, Mike Hurwitz, Jonathan Smith, and I find that the subtle impacts of hotter temperature may actually be reducing the rate of learning differently for students of varying racial and socioeconomic backgrounds. It appears that heat in the learning environment can be significantly disruptive.[17]

Comparing the same students in the same school over multiple years allows us to get as close as possible to teasing out the subtle causal effect of temperature on learning, as opposed to many other factors that may be correlated with geography. Moreover, when we look at the effect of hotter temperature on learning during school days compared to the effect on non-school days (weekends, holidays, summer vacation), we find that the effect is driven primarily by hotter school days. One interpretation is that this is consistent with much of the damage coming through disrupted class and learning time.

As with the earlier study on natural disasters and learning, the data and methods we employ reflect an important shift in how social scientists think about causality in the context of climate damages, a shift we will explore in greater detail in the following chapters.

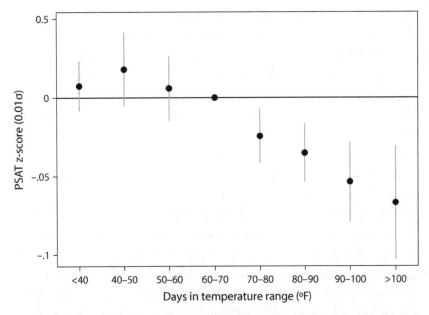

FIGURE 2.4. The relationship between temperature and learning. The figure plots the change in a student's individual PSAT score (hundredths of a standard deviation) on the number of days in the preceding year in a given temperature range, where days in the 60°F–70°F range is the baseline category. The regression includes student fixed effects and fixed effects for each combination of cohort, test date, and take number. Bars represent 95 percent confidence intervals. The sample comprises all students from the high school classes of 2001–2014 who took the PSAT more than once (approximately 21 million test scores). Source: R. Jisung Park et al., "Heat and Learning," *American Economic Journal: Economic Policy* 12, no. 2 (2020): 306–339.

The effect of hotter temperature on student achievement is not very large on average. Over the period from 1998 to 2011, we find that a school year with roughly 1°F hotter-than-average temperatures reduces the rate of learning—as evidenced by gains in PSAT performance over the course of that year—by around 1 percent for the average American secondary school student (see figure 2.4). For most individuals in most years, such an effect would be hardly noticeable.

But, because they persist over time and are spread out over many millions of students, the aggregate impacts appear to add up in economically significant ways. We estimate that, without remedial investments in adaptation, projected warming could reduce overall learning outcomes in the United States by 10 percent of the average learning achieved in a given year.

Importantly, we find that the average effect masks important differences by race and income, suggesting that heat may be far more disruptive of learning for some than others. For any given hot day, Black and Hispanic students are two to three times more adversely affected than their White or Asian counterparts.

Along with the fact that Black and Hispanic students tend to live and go to school in parts of the country where heat during the school year is more frequent, and tend to learn in classrooms that are less likely to be air-conditioned, we estimate that the disparate impacts of heat on learning may actually be contributing to racial academic achievement gaps. By our estimates, hotter temperature associated with our current climate—not from future climate change—may by itself be accountable for up to 5 percent of current Black–White achievement gaps, which are on the order of roughly three years' worth of formal schooling. Climate change could increase these gaps further.

How about students outside of the United States? The average American student experiences approximately twelve school days per year with temperatures above 32.2°C (90°F). The average Indian student experiences over 100 such days annually, suggesting that even if Indian students had similar access to cooling technologies such as air-conditioning, heat could hamper learning to a much greater degree.

In another study, economists Patrick Behrer, Josh Goodman, and I find that similar effects of heat on learning may be at play across the world.[18] Using data on standardized test scores for over sixty-eight different countries including Brazil, South Korea, and Germany, we find that hotter school days reduce the rate of learning, as measured once again by gains in average standardized test performance.

The effects of heat on learning appear to be larger for younger students than older ones. They also appear to be concentrated in low-income populations, whether that is lower-income school districts within the United States or lower-income countries across the globe. While a major drawback of this study was that, due to data limitations,

PHYSICAL VERSUS HUMAN CAPITAL   51

we were unable to include many of the world's poorest countries, particularly many parts of Sub-Saharan Africa and South Asia, the evidence nevertheless suggests that hotter temperature may already be contributing to international gaps in learning, which may themselves be driving some differences in standards of living.

A glance at the following statistics suggests that the stakes may be high. The average American experiences 44 school days per year with hot temperatures, with highs above 80°F (26.6°C). In Turkey, South Korea, and France, the number is 20, 36, and 11 school days per year, respectively. Compare that to the average student in Thailand, or Mexico, or Brazil. These students experience 203, 140, and 122 school days per year where hotter temperature may subtly but meaningfully dampen students' ability to concentrate and learn.[19] One can imagine how, over time, such subtle diminutions may add up.

These studies provide some initial clues—intimations that climate change may have a more direct and persistent influence on educational outcomes, and possibly economic growth, than previously appreciated.

They suggest that heat may already exert a subtle but profound influence on our daily lives and economic livelihoods, possibly exacerbating inequality of various forms. The adverse human impacts of heat, while small on the margin, possibly even imperceptible in any individual case, may add up in ways that are important, as if throwing sand into the gears of the economy. How we measure these often subtle effects, and whether we can be confident that they tell us something meaningful about climate change, is not obvious and is a question we will unpack further in the chapters ahead.

What this perspective also reveals are the potentially overlooked points of possible intervention. In the case of heat and learning, the evidence suggests that these unequal adverse effects are not immutable. At least in the United States, air-conditioning at school and at home appear to be effective at significantly reducing heat's effects on learning, though it may be that other as yet unmeasured investments that protect the learning environment from adverse weather impacts, such as tree cover or cool roofs or smarter school policies, may be important in reducing heat's negative effects as well. In fact, our analysis suggests that something as relatively simple as bringing students to an equal physical playing field may help close achievement gaps and protect against further exacerbation of such gaps due to climate change.

This begs the critical question of what kinds of interventions can most cost-effectively and equitably reduce climate vulnerabilities in the years to come, a multifaceted problem whose complexities we will decipher in the chapters ahead.

In this chapter, we've explored the idea that when it comes to climate change, the nonmarket damages may be nontrivial, possibly central. Human capital is but one dimension; there are likely many others. These include the deterioration of physical health and psychological well-being; degradation and possible destruction of entire ecosystems, like coral reefs and the biodiversity they encase; the destruction of sites of cultural and historical significance, like the ongoing inundation of ancestral graves in many small-island nations or the hastened subsidence of architectural treasures in low-lying cities such as Venice. As we will see in chapter 7, they may even extend to the subtle erosion of the social fabric that many of our daily economic interactions depend on, including public safety, the fair functioning of the courts, or the civility of public discourse.

3

Where There Is Smoke

IT'S HARD TO FORGET the shock of those first nights in Los Angeles. My wife and I had just moved into our new apartment. Here we were, newlyweds: me having taken up a faculty position at the University of California, Los Angeles (UCLA), she having recently received her green card. Both of us were excited to start a new life together in California.

The images of a hillside on fire—towering flames many stories high, thick black smoke plumes billowing as far as the eye can see—are seared into my memory. This was December 2017, when the Skirball fire was burning less than a few miles from UCLA.

Given a peripatetic upbringing, I thought I'd experienced the gamut of natural hazards: tornadoes in Kansas, typhoons in South Korea, blizzards in New England, and even a seasonal-affective-disorder-inducing lack of sunlight in the United Kingdom. But wildfire was a categorically new experience for me. The shock was visceral.

I received a text that evening from one of my colleagues. "Are you doing okay? It looks like there may be a mandatory evacuation order coming into effect. I'll keep you posted."

We had just begun to unpack. My wife wore a look of concern as she pulled dishes out of a cardboard box. "So . . . should we stop unpacking?" she asked. I gave her a look of bemusement. "I'm not sure. I guess we'll have to wait and see."

A mandatory evacuation order did come into effect that evening. Thankfully for us, it stopped two blocks from our apartment. Not everyone was so lucky. Thousands of our neighbors were forced to evacuate. When all was said and done, over a dozen residences were burned, many completely destroyed. Classes were canceled and exams postponed at UCLA, and nearby elementary and middle schools shut down for days.

You may already be familiar with the fact that, as climate change progresses, wildfires will likely burn hotter and occur more frequently, in some cases causing nearly complete destruction of entire towns, as in the case of the Camp Fire and the towns of Paradise and Concow, California, which lost over 95 percent of their structures and over 85 lives.[1]

You may also be thinking, "Enough with the theatrics already. I get it. Climate change is real. It's scary. We need to get our act together fast."

Fair. Climate change is happening faster than most people think. And as a grown man, I am not afraid to admit that those flames were truly terrifying. But that is not the point of the story.

The point of the story is not the flames of the raging wildfire, or the climate apocalypse that such conflagrations may portend.

The point of this particular story is the smoke. What if it was the case that, when it comes to the societal cost of wildfires today, it may actually be the smoke that causes the lion's share of overall harm due to its incremental effects on learning, productivity, and health? And that the damage from the smoke is not limited to residents of Los Angeles or California or even the American West, but extends to varying degrees across the world, affecting people as far as Boston, Chicago, Montreal, and Mexico City?

Similarly, what if it were true that the hidden costs of a warmer world may ultimately exact a social and economic toll on par with several times the annual profits of all Fortune 500 companies combined?[2] And that the subtle but pervasive health impacts of slightly hotter temperature *alone* may surpass all other natural disasters many times over, and may be exacerbating many of the forces that make modern economies more unequal?

What if it was the metaphorical smoke—not the fire—that kills more of us in the end?

The Hidden Consequences of Wildfire Smoke

In the days following the Skirball fire, I found myself grappling with the tension between system one and system two thinking. The towering flames were truly terrifying. The scene *felt* apocalyptic. I realized that I was triggered, at the level of my entire nervous system. My fight-or-flight responses were operating on hyperdrive.

Two intuitive judgments kept wanting to congeal in my psyche. The first had to do with my wife's and my personal safety, and how we might adapt to this new threat. In retrospect, my internal dialogue was embarrassingly crass, emotional, and self-centered. Something to the effect of "I wish we had known more about these wildfires *before* we moved!" and "at least we didn't buy a house yet."

The second was bigger picture, more to do with my evolving sense of how much we as a society should care about reducing emissions to combat climate change globally. But here too, the gut reaction—if I am honest with myself—involved colorful internal language. "If this is a harbinger of what's to come, perhaps Greta is right and we should all be panicking and boycotting school after all. Maybe climate change really is an existential crisis for humankind."

But my training as an applied econometrician was imploring me to look at the situation with a bit more system two thinking: both for my wife and my personal decisions—for example, whether to buy a house in that neighborhood or in Los Angeles generally—and for our evolving assessment of the seriousness of the climate problem globally.

It made me inclined to consider a range of questions: How often have wildfires of this intensity happened in our neighborhood? What does the evidence say about the likelihood of such events in the future given climate change?

In terms of my judgments about climate change writ large, a number of additional questions followed: How confident are we that climate change will increase wildfire frequency or severity? What do we know about the effectiveness of various hazard mitigation strategies?

With hindsight, I can see that this dissonance was at its peak the day after the harrowing near-evacuation. Once the flames were brought under control, life more or less went back to normal. Classes resumed, restaurants reopened, and media outlets stopped covering the story as breaking news.

I found myself sitting in the UCLA sculpture garden, enjoying lunch with new colleagues. Having endured many dreary New England winters as a graduate student, the opportunity to have lunch outdoors in November came as a delight. Except that the smoldering embers of the fire were still billowing smoke for miles, and campus was increasingly becoming blanketed in a cloud of gray—the kind of sooty ashen air that surrounds a recently doused campfire.

The Air Quality Index (AQI) that afternoon around campus was above 150, several times the level deemed safe by the Environmental Protection Agency, closer to levels typically observed in Beijing. I noticed students coughing and covering their faces on their way to and from class. Perhaps it was a placebo effect, but my eyes were itchy. The air smelled like someone had left a burning stove unattended. I found myself wondering whether we should be having lunch out here after all, or whether we should be holding final exams under these conditions. But everyone was relieved that the flames had been contained, the threat of imminent loss of life curtailed, that there were no longer any helicopters hauling giant carafes of water overhead.

So we went on with our daily lives—outdoor lunches and final exams and all. The terror of the wildfire eventually subsided. And when in February the rains came and the flowers bloomed, California felt once again like the land of golden opportunity that I had dreamed of. That is, of course, until the following autumn, when the deadliest and most destructive wildfire season on record hit the state, with over 7,500 fires burning over 1.8 million acres, covering the region in clouds of ash and smoke once more.[3]

What I didn't fully appreciate until after the fact was that the question I should have been raising more forcefully was about the smoke. That, despite prevailing social norms, I should have asked whether we should be back on campus so soon, with all this lingering smoke in the air.

Since 2005, there have been approximately 70,000 wildfires in the United States each year. Many of these, thankfully for humans, occur in remote areas mostly removed from major population centers. Still, over the past four decades, burned area from wildfires has nearly quadrupled in the United States. According to recent studies, climate change may be responsible for around half of this increase, due in part to increased fuel aridity resulting from rising temperatures and drought.[4] And because more and more Americans are moving to places that are susceptible to wildfire—to areas known as the wildland-urban interface (WUI)—the number of people and homes exposed to wildfire is increasing as well. In fact, by one estimate, the number of homes in the WUI has grown by approximately 350,000 per year since 2005.[5]

In that period, wildfires have destroyed nearly 6,000 homes and caused around twenty to thirty deaths per year.[6] It is worth repeating this number: twenty to thirty officially recorded deaths per year in the United States due to wildfires. Lest we lose sight of the global nature of the problem, catastrophic wildfires have killed hundreds in Australia, Portugal, and Greece as well.[7] But let's focus on the US numbers for a thought experiment.

More wildfire means more flames. But it also means more smoke. In some cases, the smoke can linger for far longer than the fire itself. Especially when combined with dry, stagnant air, like in Los Angeles, wildfire smoke can settle atop urban centers and mingle with air pollution from vehicles and industry for days, even weeks. In other cases, strong winds whisk the smoke away from burned areas rapidly, keeping the air in neighborhoods adjacent to the actual burn sites remarkably clear, but transporting the smoke to distant communities. Indeed, satellite imagery suggests that smoke can carry long distances: *very* long distances. Smoke plumes from fires in the Pacific Northwest can have measurable effects on air quality in New York or Boston.

Residents of the northeastern United States had an alarming glimpse of this during the summer of 2023, when wildfires in Canada created dense smoke plumes that blew over major urban areas, including New York, Chicago, and Washington, DC. The severity of the pollution was unprecedented. On June 7, New York was the most polluted major city in the world.[8] Particulate matter concentrations in New York hit 195 micrograms per cubic meter, roughly five times the national air quality standard. The last time New York City residents experienced such significant levels of smog was probably in the 1960s, before the federal Clean Air Act went into effect.[9]

According to one analysis, that day may have been the worst day of exposure to wildfire smoke for the average American since such measures started being recorded in 2006.[10] The average PM2.5 (the number of very small particles of 2.5 microns or less in diameter in the air) experienced across all Americans that day was 27.5 micrograms per cubic meter. The second worst had been 17.5 micrograms on September 13, 2020, following a record year of wildfires in the western United States.

One of the reasons why this particular episode was so record-breaking has to do with the joint spatial distribution of pollution and people. Because so many people live in the northeastern United States, and because wildfires of this intensity have historically been rare in eastern Canada, this fire managed to affect the lives of many more people than similar fires in the American West, including California, albeit in largely subtle ways.

In some cities, including New York, this episode was acute enough in its intensity to mobilize considerable public response. Governor Kathy Hochul issued a public statement recommending school districts cancel their outdoor activities. Thousands of flights were delayed or canceled, and no doubt many gatherings large and small—weddings, concerts, outdoor yoga sessions—needed to be canceled or postponed as well.

This suggests that wildfires have the potential to affect many millions of people living hundreds, even thousands of miles away. It also means that estimating the realized human exposure to wildfire smoke is challenging. It is more than simply a physics or atmospheric chemistry problem—it is also an economic and social one. Doing so requires asking many difficult-to-answer questions: How many people are exposed to the shifting smoke plumes and for how long? To what extent does being indoors protect against the adverse consequences of smoke exposure, and what proportion of people change their behaviors in response to smoke? How many are likely to be more heavily exposed because they work outdoors for a living? Who is more likely to purchase air filters for their offices and homes?

Combining state-of-the-art physical models with more granular data on social and economic outcomes, a slate of new studies tries to generate estimates of the potential societal costs of wildfire smoke, accounting for such socioeconomic nuances as much as possible. Often, they employ satellite imagery, ground measures, and atmospheric models in conjunction with statistical tools that can help isolate the causal contribution of smoke on human outcomes—tools that we will unpack in the chapters to come—to estimate this more precisely. An emerging prognosis is that climate change may lead to a manifold increase in smoke exposure across the United States and worldwide.[11] One estimate suggests that, across the United States, the number of people exposed to extreme smoke days has increased twenty-sevenfold in the past decade alone.[12]

Moreover, wildfire smoke may now be responsible for 20 percent of all PM2.5 pollution exposure in the United States, and up to 60 percent in some regions, a number that will likely grow as climate change progresses.[13]

Here's where the hidden costs become important to consider. What I didn't fully appreciate at the time of the Skirball fire was just how harmful the smoke really might be. Recent research suggests that the hidden health costs of the smoke alone may vastly exceed the direct destruction and death from US wildfires.

Despite significant progress since the passage of the Clean Air Act, air pollution remains a persistent problem in many parts of the United States. In fact, a growing body of evidence suggests that increased wildfire smoke may be responsible in part for the slowing, and in some regions stark reversal, of trends in improved air quality.

Across a range of studies, it is estimated that PM2.5 pollution is responsible for anywhere between 50,000 and 100,000 premature deaths in the United States alone.[14] Mortality from air pollution is likely many times this figure in places like India and China. Some estimates suggest that air pollution in China has resulted in reduced life expectancy of roughly five years per person: over 7 billion life-years lost, though it is important to note that, in much of the world, wildfires are not the dominant contributor to air pollution, and that there are still many open questions about the relative virulence of smoke from wildfires as opposed to other sources of particulate matter such as vehicle or factory emissions.[15]

Tallying the downstream economic and health costs of smoke exposure, economist Marshall Burke and his colleagues estimate that increased wildfire smoke may have caused between 5,000 and 15,000 additional deaths per year over the period from 2006 to 2018, mainly among the elderly. Very few of these will be officially categorized as having been caused by wildfires because they will have been the result of the cumulative influence of worsened air quality and weakened health over the course of months if not years.[16]

It is worth pausing a moment to put numbers like this into perspective. Wildfires in the United States may cause an additional 5,000 to 15,000 people to die prematurely each year. I mentioned at the outset that, according to official estimates, wildfires have killed roughly 20–30 people per year. Each of these deaths is a graphic tragedy. Families blindsided by a wall of flame in the middle of the night. Firefighters perishing in a valiant attempt to save others. But what the data suggests is that, for each of these more visceral human tragedies, another 150 to 750 may be dying a silent death of a thousand tiny cuts, sometimes in far-off cities where wildfire risk is little more than an episodic curiosity.[17] Of course, these estimates are subject to important caveats, including the fact that we do not yet know whether PM2.5 from wildfire smoke is similarly deadly as other sources of PM (as is assumed in the Burke et al. study). As a rough indication of the potential magnitude of harm, however, these figures are instructive.

The hidden consequences of wildfire smoke may cut even deeper. Research now suggests that the cumulative effect of wildfire smoke on subtle things like mental health, learning, and productivity may be economically significant as well.[18] For instance, economists Joshua Graff-Zivin and Matt Neidell found that workers on California's farms and packing facilities are 6 percent less productive on days where PM2.5 is above fifteen micrograms per cubic meter.[19]

It is worth noting that US National Ambient Air Quality Standards sets thirty-five micrograms per cubic meter as the twenty-four-hour threshold for the United States. While there isn't a one-to-one equivalence from PM2.5 to AQI (since AQI is typically a composite index of a number of pollutants including but not limited to PM2.5), the effects they document might be thought of as follows: a day with AQI above 100 (a "poor" air quality day) can reduce manufacturing productivity by 12 percent or more relative to a day with AQI around thirty (a "good" day).

Similarly, studies of Israeli students taking important exams across more polluted and less polluted days suggest that highly polluted days—for instance, a day with an AQI above 100—reduces test performance by roughly 15 percent of a standard deviation and may even affect the kinds of colleges that students are ultimately admitted to.[20]

Air pollution has also been shown to lead to higher risk of Alzheimer's and dementia over time, higher risk of low birthweight, and higher rates of hospitalizations due to asthma, suggesting that when one considers all of these and other consequences of air pollution on morbidity and quality of life the aggregate impact may be even larger than the mortality estimates mentioned above.[21] Once again, these findings are not merely correlational. They represent a growing body of environmental economic research that uses natural experiments—for instance, random changes in pollution exposure arising from differences in wind direction over time—to carefully disentangle the causal footprint of pollution from other potentially correlated factors, like income or occupation.

While the evidentiary base continues to evolve, it is becoming increasingly difficult to dismiss the idea that, when it comes to wildfires, the social costs of the smoke could potentially be on par with, or larger than, the direct damages from the flames. The case of wildfire smoke and its subtle but pervasive human costs may hold important lessons for how we think about the threat of climate change.

Disasters Large and Small

There is no shortage of apocalyptic imagery surrounding climate disasters, whether in the form of "biblical floods," inundated cities, or towns engulfed by a wall of flame. Sometimes, it may be what jolts us out of complacency to question the status quo, as was the case for Greta Thunberg.

Relatively less attention is paid to the smaller costs. These are the billions of lower-grade climate burns that, at least according to the data, appear to have cumulatively meaningful consequences even in today's climate, not to mention the hotter one that we have in store.

I have tried to make the case that these hidden costs are worth paying closer attention to. Seeing these costs clearly requires, among other things, a willingness to check our gut impulses and to think statistically. And, as we will discuss, considering the hidden costs may be important in our search for policy solutions. For instance, might there be ways to manage wildfires that better balance protection of homes—which is typically the focus of fire crews today—and reduction of downwind pollution exposure?

This is by no means a novel insight; it is merely an application of an old idea to the particularly bias-prone and pressing societal issue of climate change. As Stephen Pinker puts it in his book about the steady decline of violence across human history: "Our cognitive faculties predispose us to believe that we live in violent times, especially when they are stoked by media that follow the watchword 'If it bleeds, it leads.' The human mind tends to estimate the probability of an event from the ease with which it can recall examples, and scenes of carnage are more likely to be beamed into our homes and burned into our memories than footage of people dying of old age."[22]

While it is important to recognize the record-breaking increase in extreme climate events, there is a risk that, when combined with the climate calamity framing, the message becomes one about a world that is facing imminent collapse. As we will discuss in the coming chapters, such apocalyptic framing is problematic on at least two dimensions that are relevant for how we make important decisions moving forward: namely, it casts mitigation as a binary decision as opposed to a spectrum, and it potentially distracts from the real adaptation priorities, especially given the many social determinants of climate vulnerability.

So far, we've explored the possibility that the hidden costs of climate change add up by way of two case studies. The human capital example suggests that the nonmarket damages from climate change may often be of a similar order of magnitude as the easier to quantify market damages. The smoke and flames example suggests that sometimes the damages may accumulate in places and contexts far removed from the source of the more salient climate risk.[23] Next, we will explore the possibility that incrementally hotter temperatures may, through their largely unnoticed effects on a wide range of social and economic variables, actually exert a much larger influence on economic development and human well-being than many may recognize.

PART II

IN THE 1999 MOVIE *Office Space*, a ragtag team of cubicle workers engage in a hilarious get-rich-quick scheme. Fed up with their dead-end jobs, Peter and his associates devise a plan to shave fractions of a penny off each financial transaction that their company engages in, sending these illicit shavings into their own hidden savings account. Their hope is that such small pilfering will go unnoticed but will add up over time. Their master plan is to let this account accumulate silently and slowly, just enough so that they would have a nice cushion by the time they retire.

Of course, the comedic element is that they meant to shave *fractions* of a penny, but mistakenly shaved pennies per transaction. And so, the illegal earnings balloon out of control far more quickly than they anticipated—to over $300,000 in a matter of days—forcing Peter to come clean about their wrongdoing.

Could something similar be at play with climate change? For the many trillions of economic transactions that occur each day, might it be as if an invisible collector is shaving pennies to the dollar off the value of each transaction?

The answer depends in part on how sensitive the gears of the modern economy are to incremental increases in temperatures, including the potential effects on the billions of workers, consumers, and decision-makers who make it run.

Part II of the book explores in greater detail the role of temperature in affecting human health, productivity, and well-being broadly construed. We will explore the science of how heat affects health, and how new data and statistical tools help illuminate many hitherto hidden harms. We will examine the effects of hotter temperature on human performance—from physical performance of athletes to cognitive performance of students—and whether they may be relevant in thinking about the relationship between climate and economic development. We will also explore whether climate change may influence our emotions, our social interactions, and the safety of the social environment in which we go about our daily lives.

Doing so requires a bit of epistemological inquiry—a sensitivity to the process of scientific discovery, and a willingness to remain curious about how we know what we know and how confident we can be that correlation isn't being mistaken for causation.

4

Anecdotes, Data, and the Problem of Causality

JUNE 5, 2014, is a day that LeBron James probably wants to forget. In game 1 of the 2014 National Basketball Association (NBA) Finals between the Miami Heat and the San Antonio Spurs, an electrical outage in the San Antonio arena caused the air-conditioning to malfunction around halftime. Ambient temperature during the game hovered around 85°F (29°C). Inside, the fans and the players were beginning to feel the heat. ABC's Doris Burke reported during the game that the temperature in the arena had exceeded 90°F (32.2°C).

With around nine minutes left in the fourth quarter and the Heat leading 85 to 79, James asked for a breather—an unusually early point in the game for the league MVP to take a seat on the bench. A six-point lead was by no means a comfortable margin, not when playing on the road against the four-time NBA champion Spurs, led by future Hall of Famers Tim Duncan, Tony Parker, and coach Greg Popovich. The stakes of this first match were especially high, given that over 75 percent of the time, teams that win game 1 of the NBA Finals go on to win the championship outright.

With LeBron out of commission, the Spurs went on a furious run, making the score 90 (Miami) to 94 (San Antonio) with less than five minutes left to play. To anyone watching the game live, it was clear that the momentum had swung in the home team's favor. The crowd was electric. The rest of the Heat players looked stunned. After sitting with ice packs on his neck to try to cool off, LeBron attempted to return. But his body had other plans. With 3:59 remaining on the clock, an eternity in professional basketball, he left the game for good. San Antonio finished the game on a 16–3 run. Final score: Miami 95, San Antonio 105.

The Spurs would go on to win the series four games to two, clinching their fifth NBA championship. That summer, LeBron James would leave the Heat as a free agent, much to the dismay of thousands of Miami sports fans.

Was temperature a determining factor? Given the drama and the stakes, fans and commentators were clamoring to know.

"[The heat] was significant. It was definitely a factor," San Antonio's Tim Duncan told ABC after the game. "I don't know about what happened to LeBron, but all of us feeling the heat were dehydrated."[1]

As ABC's Mike Breen reported midgame, "If you're watching from home, I don't know if you've noticed, fans are fanning themselves." His colleague Mark Jackson noted that the temperature could start to factor strategically: "This is probably going to take more out of the players. As a coach, you need to make sure you're monitoring this."

After the game, social media posts speculated that the Spurs' coach Greg Popovich may have secretly manipulated the building's air-conditioning to give his possibly better acclimated players another home-court advantage.[2]

We will never know what would have happened had it not been so hot in the arena that night. Would LeBron have held out for a few more minutes, giving his team the edge? After all, this was the man who was widely considered one of the most mentally and physically gifted athletes of all time, who had led his team to the championship both of the previous two seasons.[3] And if the Heat had won the series opener, would they have gone on to win the championship once more? If so, would LeBron have stayed in Miami?

Maybe. Maybe not. Without the benefit of what scientists call a *counterfactual*—that is, the same game played with the same players and under the same circumstances except for the unusually hot temperature—it is impossible to tell.

Magic Johnson, Larry Bird, and the "Heat Game"

This was a rare event, as most professional basketball games are played indoors with air-conditioning, but not one without precedent. Thirty years prior, a similar scenario unfolded in Boston Garden.

In June 1984, Magic Johnson and the Los Angeles Lakers faced off against Larry Bird and the Boston Celtics in game 5 of one of the most celebrated matchups in NBA Finals history. Affectionately referred to by fans as "The Heat Game," the game took place amid a then extremely unusual June heat wave in Boston.

This time, rather than a malfunctioning system, Boston simply didn't have air-conditioning installed. Not surprising given New England's relatively cool climate, and if one considers that fewer than 50 percent of American homes had central air as of 1984.[4] According to news reports, it was anywhere between 97°F (36°C) and 115°F (46°C) in the arena during the game.

Notably, the Los Angeles Lakers were used to playing with air-conditioning for their own home games. "The Lakers forum was air-conditioned and cool. We had the ratty barn. In the heat of the day. It was muggy, it was stifling, and the Lakers just weren't used to that," recalls the Celtics' Cedric Maxwell.[5]

"It was like a sauna in our locker room," said Magic Johnson of the Boston arena. "We sweat like it was pourin' down on us. Kareem was really suffering." In footage of the match, the 37-year-old Kareem Abdul-Jabbar (another future Hall-of-Famer) can be seen sitting on the bench with an oxygen mask over his face, in a desperate attempt to stay in the game.

"Even the refs were keeling over," recalls one player. It was so hot that referee Hugh Evans had to stop officiating at halftime due to dehydration.

And then there were the fans. As *Boston Globe* columnist Bob Ryan writes: "There were some hot nights in that old building over the years, but there was never one like the evening of June 8, 1984. The male fans wore shorts and short-sleeved shirts. The women wore, well, as little as possible. Halter tops proliferated. There was never a day or evening in the long history of that building when there was so much exposed skin."[6] With the series tied, this too was a pivotal game. In previous such game 5 matchups, the winner went on to win the NBA championship 70 percent of the time.[7]

After a chaotic match, the Lakers lost 121 to 103. While Magic Johnson and company managed to bounce back and win game 6 in Los Angeles, they ended up losing the series to Boston in seven games.

Boston had been the clear underdog going into the series. With the Lakers coming off two recent championship runs and buoyed by an ascendant Magic Johnson entering his prime, there had been growing talk of a "Laker Dynasty." After all, these were the "Showtime Lakers" of legend, who would go on to win a total of five NBA championships in the span of nine years.

"It was a championship that I really felt that we stole. They were the better team," recalls Celtics guard and future NBA coach Danny Ainge. "They were more talented than we were. But I think we just wanted it a little bit more," said the Celtics' Cedric Maxwell after the final game.[8]

These are just two games. Two games out of thousands, the vast majority of which occur without anyone—players, coaches, referees, fans— thinking twice about the temperature. But given what transpired, and the high stakes, they do make one wonder: How much does temperature matter for human performance?

Trying to answer this question definitively would seem a fool's errand. In the case of these specific professional basketball games, the list of potential confounding factors is long. There is home-court advantage to consider, jet lag, the potential role of sleep or food the night before. There are players' myriad interactions with each other on the court and with family members, friends, and fans off it.

Though our minds may have evolved to think in stories, the difference between generalizable scientific fact and plausible-sounding fiction can often be deceptively subtle. So to attempt a rigorous answer will require more than a few anecdotes. And, as we will see, even having access to controlled environments in the laboratory or the statistical controls provided by having many more data points may not be sufficient, either.

Hot Temperature and High-Stakes Exams

Back to June 2014. As Lebron James was shaking off his disappointment at losing in the NBA Finals, a thousand miles to the northeast, high school seniors in New York City were getting ready for their own high-stakes battles. Battles, not on the basketball court but with the enduring gatekeeper of higher education: the standardized exam.

Each June, students in New York City public schools take a series of exams known as the Regents exams. They are, in effect, exit exams: prerequisites for graduating with a high school diploma, and in many cases for having the chance to continue on to further educational opportunity in college.

For some students—like those attending prestigious magnet schools the likes of Bronx Science or Stuyvesant Academy—these exams are mostly perfunctory endeavors, maybe even a bit of a nuisance in the way of studying for the more challenging Advanced Placement (AP) or SAT exams. For many others, indeed the vast majority of a sprawling public school system serving over 1 million students, these exams can mean the difference between a shot at entering the knowledge economy and a life spent navigating a maze of minimum wage jobs, for-profit colleges, and the criminal justice system.

Whereas graduation rates among the best schools consistently top 95 percent, with many students going on to prestigious Ivy League universities, the average graduation rate for the system as a whole has hovered in the 60–80 percent range over the past two decades, with some schools and districts averaging on-time graduation rates below 40 percent. That means that, for some schools, six out of ten high school seniors either repeated a grade or dropped out before receiving a diploma. The vast majority of students come from disadvantaged backgrounds. Black and Hispanic students comprise over 70 percent of the student body, and approximately 75 percent of the district's students received federally subsidized school lunches as of the mid-2010s.[9]

As anyone who has taken the New York City subway in the summer knows all too well, the Big Apple can get pretty stifling. While June is still relatively early in the season, the mercury regularly reaches the high 80s, sometimes above 90°F (32.2°C). With so much asphalt and concrete, parts of the city can experience intense urban heat island effects locally.

The following are temperatures as measured in the weather station in Central Park for the two calendar weeks that span the Regents exam period. Tuesday, June 17 was especially hot, with temperatures in the mid-80s (°F) at 6:00 A.M. and reaching close to 90°F by noon. Not much respite that evening or the following day, with highs back up to 88°F on Wednesday. But subsequently the temperature cooled off, only reaching the low 70s (°F) on Thursday. The following week, while things remained cooler on Monday, temperatures crept up again into the mid-80s by Wednesday.

In order to minimize the risk of cheating, the timing of these exams is strictly regulated and harmonized across the district. So even if students were able to anticipate hot exam conditions, they had effectively no way of rescheduling or devising alternative testing arrangements. This ends up being important from the standpoint of data analysis and causal inference, which we will return to later.

Figure 4.1 shows the schedule for the Regents exams administered in June 2014. On Tuesday, June 17, the US History and Government exam was administered at 9:15 A.M., followed at 1:15 P.M. by the test for Living Environment. On Wednesday, Global History and Geography was administered in the morning slot and Algebra II or Trigonometry in the afternoon, and so on and so forth over a roughly two-week period.[10]

Recall the temperature on the night of LeBron's fateful game against the San Antonio Spurs. There, outdoor temperatures were in the mid-80s, and reported temperatures indoors were in the 90s. One might easily presume that as the most resource-intensive school district in the country, located in the economic powerhouse that is New York City, all of the city's schools would be fully air-conditioned. As it turns out, fewer than 60 percent of public school buildings were reported as having fully operational air-conditioning systems at the time.[11]

Could hotter temperatures have affected the performance of students taking their exams that year? Unlike the NBA, we do not have play-by-play accounts or video recordings of the events. We do, however, have administrative data. As part of its internal recordkeeping, New York

ANECDOTES, DATA, AND THE PROBLEM OF CAUSALITY 73

The University of the State of New York
THE STATE EDUCATION DEPARTMENT
Office of State Assessment
Albany, New York 12234

Examination Schedule: June <u>2014</u>

***UPDATED 1/27/14** – This schedule supersedes any previously released schedule.*

Students must verify with their schools the exact times that they are to report for their State examinations.

June 3 TUESDAY	June 17 TUESDAY	June 18 WEDNESDAY	June 19 THURSDAY	June 20 FRIDAY	June 23° MONDAY	June 24 TUESDAY	June 25 WEDNESDAY	June 26 THURSDAY
9:15 a.m.	9:15 a.m.	9:15 a.m.	9:15 a.m.	9:15 a.m.	9:15 a.m.	9:15 a.m.	9:15 a.m.	**RATING DAY**
RE in Algebra I (Common Core) ♦	RE in U.S. History & Government	RE in Global History & Geography	Comprehensive English	Integrated Algebra	RCT in Global Studies*	Physical Setting/ Chemistry RCT in Science*	RCT in Writing	
1:15 p.m.	1:15 p.m.	1:15 p.m.	1:15 p.m.	1:15 p.m.	1:15 p.m.	1:15 p.m.	1:15 p.m.	**Uniform Admission Deadlines**
RE in English Language Arts (Common Core) ♦	Living Environment	*Algebra 2/ Trigonometry*	Physical Setting/ Earth Science	Geometry *Physical Setting/ Physics*	RCT in Reading	RCT in U.S. History & Government*	RCT in Mathematics*	Morning Examinations 10:00 a.m. Afternoon Examinations 2:00 p.m.

° Suggested date for administering locally developed tests aligned to the Checkpoint A and Checkpoint B learning standards for languages other than English.

* Available in Restricted Form only. Each copy of a restricted test is numbered and sealed in its own envelope and must be returned, whether used or unused, to the Department at the end of the examination period.

♦ Conversion Charts for these exams will be available no later than June 26, 2014.

FIGURE 4.1. The exam schedule for New York State Regents exams. *Source*: New York State Education Department.

City keeps track of its students' performance on these and other important exams year after year. Perhaps this data, a form of big data (in that it was not originally collected for the purposes of research), could help us answer this question. We will come back to this data and the promise and perils it presents momentarily.

For now, let us consider the hypothetical situation faced by a demographically representative student.[12] Take Sofia, a 17-year-old woman from Brooklyn, who was finishing up her senior year of high school. Sofia's grades are on the border of OK and good. She finds Trigonometry and Biology to be interesting and is hopeful that she might even score high enough to receive a proficiency rating (which requires a grade of 85 out of 100 or higher). But US History is really not her thing, nor is Earth Science. She would be happy simply to pass her exams for either subject with a 65 or higher.

Suppose Sofia did in fact eke out an 85 on Trigonometry, and even managed to pass Biology with a 71. But her US History score was a 63—a failing grade. Based on her previous year's exam scores, she would need to retake the US History exam in August in order to graduate with a high school diploma.

Sofia is devastated first and foremost not to be able to walk in the graduation ceremony with her friends later in the month. But she thinks that, with a little bit of studying over the summer, she can do well enough on the US History exam to receive her diploma in the end. She hopes that this might even allow her to apply to one of the local City University of New York college programs, which just might give her a shot at being an accountant at one of the major firms in Manhattan someday.

But neither of her parents went to college and right now she feels pressure to help them pay the bills and care for her two younger siblings. So she prioritizes finding a full-time summer job as a sales cashier at the local bodega. The months go by, and by the time August rolls around, her memory of the Gettysburg Address and other important events in US History have faded even further despite her ambitions of reviewing her notes in the evenings after work. Which president signed into law the Indian Removal Act again? Where did Shay's Rebellion originate? Disappointed in herself and unsure of whether it is worth it at all, she decides not to retake the exam. Just like that, she becomes branded as a high school dropout.

Of course, this is a fictitious account, embellished to help illustrate a point. But the data suggests this is by no means an uncommon experience. The median Regents score for Hispanic girls was 66 out of 100, just above the passing threshold for most subjects. In the administrative data, which is stripped of names and personal addresses to preserve students' privacy but retains key demographic information such as race and ethnicity, we observe that many minority students do not to take their make-up exams in August despite the fact that it represents their best shot at securing a valuable high school diploma.[13] Research now suggests that this is a general phenomenon across the country. Black and Spanish students are significantly less likely to retake important exams like the Regents or the SAT, even though they are likely to benefit most from doing so. In fact, underrepresented minorities are nine percentage points less likely to retake an SAT, which can only be partially explained by differences in household income, even though retaking often leads to a substantial gain in scores, particularly for underrepresented minorities and low-income students.[14]

More pertinently for us, this story simultaneously illustrates the theoretical possibility that a student's test-taking conditions could have ripple effects for the rest of their life, while also illuminating the near impossibility of knowing for sure in the case of any given individual.

Did temperature on the day of the US History exam make a meaningful difference for Sofia? Could temperature in schools be posing an impediment to public education?

For many years, teachers in New York City schools appear to have thought so. As early as 1995, the *New York Times* has been reporting on sweltering classroom conditions, particularly in the months of June and September. As one schoolteacher put it back in 2005 in response to a June heat wave, "I have children who are passing out. . . . This is not education."[15]

Teachers' unions have petitioned for classroom air-conditioning on multiple occasions. As the president of the United Federation of teachers noted, "It's inhumane to subject kids and adults to schooling in this kind of heat. . . . If this doesn't convince people that we need to air condition schools, then I don't know what will."[16]

But because air-conditioning is not required by the state, and is non-trivially expensive to install and maintain, such complaints often came to a dead end. In 2012, New York City mayor Michael Bloomberg had this to say in response to reporters' questions about school air-conditioning: "Life is full of challenges, and we don't have everything we want. We can't afford everything we want. And I suspect if you talk to everyone in this room, not one of them went to a school where they had air-conditioning."[17]

We can debate whether Mr. Bloomberg's delivery could have exhibited more empathy. But the stance is emblematic of a real challenge in the policy world. Without looking at the data systematically, it is hard to know whether there is a meaningful effect of temperature on student performance, and whether doing something like adding air-conditioning or even changing exam timing would help enough to justify the added costs—administrative and financial.

These scenarios raise some important questions. Questions that, to better understand the full social and economic consequences of a hotter climate, we may need to know the answers to.

What is the relationship between temperature and human performance? What are the implications of a hotter climate for economic productivity and growth?

These examples also illustrate some of the many social details that might matter, and that may not be easily addressed by the physical sciences alone.

How much of the effect of temperature on human performance, if there is one, might be due to physiological effects on body and brain chemistry, as opposed to psychological factors, like effort or irritability? And if the latter plays a role, how should we as a society interpret the realized consequences? If some groups experience more extreme temperatures more regularly, what does this imply for equity and fairness, and how best should decision-makers respond?

In the case of Magic Johnson and the Lakers' withering performance in game 5, was it because the Hollywood Lakers were somehow more "mentally soft" than their blue-collar Celtic counterparts? As insensitive as some of these suggestions may seem today, they appear to have been entertained broadly at the time of the match, which was held in the early 1980s.

Bob Ryan of the *Boston Globe* again:

There was one player who applied mind over matter better than everyone else, one player who not only overcame the circumstances to play a good game of basketball, but who so took to the conditions that he played one of the great games of his life. "I play in this stuff all the time back home," sneered Larry Bird. "It's like this all summer."

The other great force that night was the crowd, which turned what could have been a negative into a complete positive by celebrating the absurd conditions. Rather than bemoaning the heat, those savvy people celebrated it, realizing that the Lakers were feeling sorry for themselves because they were used to the creature comforts of the palatial Forum.

Here was the message: Watching a game in an old, cramped, steamy building and sitting on those hard seats, why, that's what we do here in New England. We don't need your cushioned seats and we don't need no stinkin' air-conditioning. We leave that stuff to you West Coast wusses. And, by the way, your team is soft.[18]

Few others could have put it quite so vociferously. But media headlines from after the game—including "Tragic Magic" on the front page of the *Los Angeles Times*—suggest that this was a dominant social narrative at the time.

The tendency to project cultural stereotypes and psychological biases is not limited to commentary of sporting events. A quick glance at news headlines would suggest that it applies to a wide range of issues—climate change included. One of our objectives in the chapters that follow is to seek a more dispassionate account in the data: to employ the scientific method in the service of a clearer discernment between fact and opinion regarding the question of how hotter temperatures might affect social outcomes.

Less culturally charged may be the practical question of whether we expect temperature to matter for purely cognitive as opposed to physical activities. In the case of the NBA players, we are talking about extremely strenuous activity.[19] Do we expect the same impacts of heat on human performance to occur for less obviously physically strenuous activities? Activities like farming, construction work, or lifting boxes in a warehouse? What about mostly cognitive activities like taking a math exam, filling out an order form, or reading a court document?

How about for the elderly and the infirm, what do hotter temperatures mean for those whose physical systems are weakened to begin with? That heat can be deadly at extreme levels is obvious, but an important question worth asking ourselves is whether heat can have significant health consequences at more moderate doses.

Finally, there are important questions around adaptation and distributional equity. If hotter temperatures turn out to systematically affect human cognitive and physical performance, what might this mean for how we understand the challenges of development in the tropics? In places like Ghana, Thailand, or Venezuela? How might climate change then affect the economic prospects of the world's poor?

Attempting to answer these questions will require more than armchair theorizing and a handful of anecdotes. An important development in the study of climate change (and of economically important phenomena generally) has been the rise of big data and the refinement of the statistical tools leveraged in pursuit of such insights.

The need for more refined tools arises in part out of a desire to leverage the power of the scientific method, in particular the power of experimentation, but to do so in more policy- or decision-relevant settings. It is one thing if the question at hand is simply a matter of ergonomics: Does temperature affect human physiology? Such questions can be answered by recruiting subjects and running experiments in a lab, for instance asking individuals to perform defined tasks under different temperature treatments. However, if the questions we are interested in include how climate affects health, productivity, and livelihoods in the real world, such experiments may be of somewhat limited value. All the more so if what we are interested in are the effects of climate *change*, and how such effects might vary across different levels of development or adaptation capacity.

5

How Heat Hurts

TIME FOR A POP QUIZ.

> Question 1. Which of the following causes more deaths per year
> in the United States?
> a. Shark attacks
> b. Bear attacks
> c. Plane crashes
> d. Car accidents

The answer is d. Being attacked by a shark or a bear is, of course, an incredibly gruesome way to die. As is dying in a plane crash. There are people who refuse to dip their toe in the ocean because they are terrified of sharks. Up to 40 percent of Americans express aviophobia, or fear of flying.[1] But most of us do not think twice about getting into a car or crossing a crosswalk.

Now, let us consider the same question, just with different options.

> Question 2. Which of the following causes more deaths per year in
> the United States?
> a. Lightning strikes
> b. Floods
> c. Hurricanes and tornadoes
> d. Car accidents

The answer is still d. Traffic accidents are still overwhelmingly more deadly than lightning strikes, floods, hurricanes, or tornadoes. In fact, they are more deadly as a group than all of the natural hazards listed here *combined*. By perhaps two orders of magnitude.

In 2019, approximately 35,000 people lost their lives to car accidents in the United States.[2] Compare this to the number of Americans who died in 2019 from hurricanes and tornadoes (42), floods (92), and lightning strikes (20); that's 154 in total.[3]

Last question.

> Question 3. If you had to order the following causes of death in terms of the number of people who die each year *globally*, how would you rank them?
> a. Hotter temperature
> b. Heart disease
> c. All cancers
> d. Car accidents
> e. AIDS

We'll see how we did in a moment. For now, we will note that, globally, 1.35 million people are estimated to be killed on roadways each year.[4]

Death in the Fields

When Mr. Jaime Nuño-Sanchez went to work on September 21, 2015, he probably didn't realize it would be for the last time. A forty-eight-year-old father of three, Mr. Nuño-Sanchez arrived early in the morning to pick lemons at a citrus grove in Southern California's Coachella Valley. Having worked the fields for over thirty years, Jaime was no stranger to the summer heat, nor to the intensity of manual agricultural labor. Over the course of an average week, a picker like Jaime may harvest as many as 12,000 lemons, often working shifts that started at 5:00 A.M. and ended around 7:00 or 8:00 P.M.[5]

It had been months since it had last rained in the region, and temperatures that week regularly reached the upper 90s (°F). At one point during his shift, Jaime sat down in a shaded area, saying he didn't feel well. Minutes later, he collapsed. A fellow farmworker called 911. But by the time the paramedics arrived, it was too late. Jaime Nuño-Sanchez died in the field at 12:35 P.M.

The temperature at the time was around 90°F (32.2°C), with higher humidity than normal, making for a heat index—which is one way to capture the additional effects of humidity—of over 100°F (37.8°C).

Upon autopsy, the county coroner attributed Mr. Nuño-Sanchez's death to cardiovascular disease.

Several of Mr. Nuño-Sanchez's coworkers said he had vomited and complained of headache prior to collapsing, which might be interpreted as signs of heat exhaustion.[6] But heat illness is notoriously difficult to diagnose definitively. In the official records, Jaime's death will be recorded as another death attributable to heart disease, the leading cause of death in most countries, including the United States.

"I am not able to conclude from the available records that the episode of illness suffered by this employee was caused or likely exacerbated by a heat-related illness, or other work-related condition," is a response typical of doctors working on cases like this one.

Jaime's case is not unique. Over 200 workers in agriculture, forestry, and fishing died on the job in the United States in 2021, according to official records provided by the Bureau of Labor Statistics.[7] If one includes more populous—and baseline dangerous—industries such as construction, mining, manufacturing, and transportation, the number of workplace fatalities balloons quickly to over 5,000 per year in the United States.[8] How many of these were attributable in whole or in part to extreme heat has long been a contentious question. At least according to official records, forty-three work-related deaths in 2021 were attributed to temperature extremes.[9]

It's worth taking a moment to think about the coroner's conundrum. How might a physician discern whether a case of heart failure was the result of extra exertion in the heat, as opposed to something else that happened that day? What to make of underlying health conditions, such as high cholesterol or family history of heart disease, or the potential role of worksite hazards such as chemical exposure or operation of heavy machinery?

As the planet warms, more and more people are waking up to the potential health consequences of hotter temperatures, including in many parts of the world like the United Kingdom, the Netherlands, and Canada, where heat exposure has not typically been a concern. For reference, weather in the United Kingdom is usually so mild that, up until recently, the average Brit experienced fewer than one or two days with temperatures above 90°F (32.2°C) per year. Perhaps not surprisingly, fewer than 5 percent of British households report having air-conditioning.[10]

It is also rising to the fore as a major public concern in much hotter, poorer parts of the world like India and Thailand, where there are usually more than 100 such days each year and rapid industrialization and urbanization have simultaneously raised incomes and generated additional public health challenges. India is home to some of the world's hottest cities. There, air-conditioning usage is similarly low despite the heat, with estimates suggesting fewer than 10 percent of Indian households having air-conditioning.[11]

And yet, up until very recently, many official statistics appear to have provided us with a seriously misleading indication of just how deadly hotter temperatures can be.

Accounting for Heat and Health

Let us ask a seemingly simple statistical question: How many people die due to heat in a given year?

Before we rush to look it up online, let us play the thought experiment of having to construct an estimate ourselves, based on some form of real-world data, as detectives of sorts in a case of mortality attribution. The challenges faced and overcome through this experiment will likely inform a more forward-looking inquiry of how many people we expect to die due to hotter temperatures from climate change if we do not reduce emissions significantly.

One approach would be to count death certificates. Let's call this the *medical classification approach*. The Centers for Disease Control and Prevention (CDC) reports that in the United States between 1999 and 2010 there were 8,081 deaths due to heat. This amounts to approximately 700 to 800 per year. How does the CDC arrive at this number? As with other causes of mortality, the CDC uses official causes of death as recorded on these certificates. If you go to their website, you can observe the number of deaths by leading cause, such as heart disease, diabetes, cancer, or traffic accidents.

These represent tallies of lives lost where the official certificate lists exposure to excessive natural heat as either the "underlying cause" or a "contributing factor," or both. Of the 8,081 heat-related deaths, the CDC notes that 72 percent of them listed heat as an underlying cause, whereas the remaining 28 percent were cases where heat was a contributing factor.[12] Interestingly, an overwhelming majority of these deaths also list cardiovascular conditions as an underlying cause.

Recall the conundrum facing the physician from earlier. In how many cases can one definitively conclude that a given death was due to extreme temperature, particularly when presented with the body of the decedent and little else? In the case of Jaime Nuño-Sanchez, what additional evidence would the medical examiner have needed to definitively conclude that heat was a contributing cause?

The CDC suggests that the classification method may lead to an undercount for these reasons: "because heat-related illnesses can cause a variety of symptoms and worsen underlying medical conditions, the cause of death is frequently misdiagnosed. Therefore, estimates of the number of heat-related deaths are lower than the actual number."[13]

But how much lower? Here, an undercount of a few percentage points, even fifty percentage points, might not be such a big deal. After all, whether heat kills 800 people in the United States per year or 1,200 people per year, while a nontrivial difference, may be small potatoes relative to such drivers of mortality as suicide, hypertension, or Parkinson's disease, which took the lives of over 47,000, 42,000, and 38,000 Americans in 2021, or, for that matter, given the number of traffic fatalities (35,000).[14] If what we cared about was maximizing bang-for-buck in public health investments, perhaps there are better uses of our time and energy.

Emerging evidence suggests that counting based on official classifications may be off by an order of magnitude, possibly more, and that climate change may be adding to this burden faster than we think.

Matters of Measurement

How we measure climate impacts matters, and there is reason to believe that our traditional tools may not be well-suited for understanding the true impacts of climate change. We will use the effects of heat on human health as a gateway to grasping the magnitude of the possible mischaracterization, but as we will see the general pattern may apply in other contexts as well.

One alternative to the classification-based approach might be to look at major heat waves and estimate how many excess deaths there were relative to the average death rate in that area for a similar period of the year. In other words, within a given city or country, during an unusual extreme heat event, how many more people died—of any cause—than usual? Let's call this the *heat wave case study* approach.

By these measures, the tally grows substantially, and points to the possibility that many deaths not attributed to hotter temperatures may nonetheless have been triggered by a heat event.

In the United States, the infamous Chicago heat wave of 1995 is estimated to have caused around 700 excess deaths in that city alone. As noted by Eric Klinenberg in his book chronicling the event and its aftermath, many of those deaths were not assigned official causes because individuals were buried without proper autopsies due to an overwhelmed health system.[15] Similar patterns have been observed in the context of other extreme heat events, including heat waves in Kansas City and St. Louis in 1980, Philadelphia in 1993, and California in 2006.[16]

The heat wave approach, while an improvement over relying solely on autopsy reports, nevertheless faces significant limitations. One problem is that there are limits to generalization—a problem of external validity, which refers to the appropriateness of taking findings from one context and applying them in another. Because heat waves are only loosely defined, cities and countries have developed their own definitions that have yet to be standardized. It is therefore difficult to get a standardized estimate of the percentage increase in mortality associated with a *heat wave* because studies of heat waves vary substantially in the definition used.

If the damages from heat turn out to be convex—that is, increasing in severity of heat event—using estimates of mortality responses based on heat waves to form projections may overstate the effect. This would be akin to predicting how much more income you'd earn if you switched from flipping burgers to playing poker for a living by simply looking at the earnings of recent World Series of Poker champions.

If, on the other hand, salient heat waves lead people to engage in avoidance behaviors, then extrapolating such estimates may understate the overall deadliness of hotter temperature. Indeed, only looking at heat waves may lead to an undercount of the overall effects of hotter temperature, if it turns out that less extreme days in the 80s and 90s (°F), for instance, also increase mortality and do not register in people's minds as being dangerous enough to change behavior. Maybe most poker players earn the majority of their income through side hustles, not from major tournament winnings.

A second problem is what is called in the literature *forward mortality displacement*, sometimes unfortunately referred to as *harvesting*. This is a problem where the number of deaths immediately caused by a period of high temperatures is at least partially offset by a reduction in the number of deaths in the period immediately following the hot day or days. In other words, heat waves affect individuals who were already very sick and would have died in the near term regardless of the heat. Not accounting for the forward-displacement effect may overstate the effect of heat waves on mortality.

Third, a big scientific limitation until relatively recently was the inability to incorporate adaptation. When the temperature rises, we'd expect people to respond however they can, by refraining from outdoor activity, turning on a fan, or, if available, running the air conditioner. This means that heat–mortality relationships, particularly those documented in relatively affluent settings, may vastly understate the health risks of heat in less affluent settings because many individuals may be adapting, though at a hidden cost. In the extreme, individuals might adapt thoroughly enough that extreme temperatures would have no discernible impact on measured health outcomes, and analyses that focus only on the dose–response relationship between health and temperature would erroneously conclude that the human health burden of climate change is negligible.[17]

Moreover, if what we care about are the total societal costs of a hotter climate, we may want to know the magnitude of the resources expended to prevent the deaths that might have occurred but didn't, as well as the extent to which the heat–mortality relationship varies by average resource endowments. Not accounting for adaptation becomes especially problematic if we have reason to believe that the ability to adapt may be unequally distributed across the population. In some cases, adaptation may be positively correlated with income, such that richer people are better adapted to heat. In others, the reverse may be true due to other factors, like the age of housing stock in many richer, cooler European cities.[18]

Here is a quick thought experiment. Suppose there are 1 million people in a city. Suppose 90 percent of them have air-conditioning, reliable access to water, and social networks where family and friends will check up on them during extreme weather events. Suppose 10 percent don't have one or all these factors. Let's call that 10 percent the *disadvantaged minority*. Now suppose that, in this city, on average 1,000 people die each year due to any cause. It could be natural deaths of old age, traffic accidents, cancer, and so on. Say we looked at the excess mortality in this city in a year with an unusual heat wave, and the increase in mortality was 20 additional deaths on top of the usual 1,000. That amounts to an increase in annual mortality of 2 percent (1,020 divided by 1,000), which might not seem like a lot.

Suppose however that upon closer inspection we found that 19 of the 20 additional deaths were among the disadvantaged minority. The implication would be that the increase in mortality risk for that group is much higher. Even if we assume conservatively that their baseline mortality rate is twice as high to begin with (perhaps due to worse underlying health conditions, greater prevalence of smoking, harsher working conditions, etc.), this translates into an increase in risk of approximately 9.5 percent, as opposed to 2 percent for the city as a whole. In fact, the implication would be that heat increases mortality risk by 9.5 percent for the disadvantaged, and 0.25 percent for everyone else, a more than 40-fold difference.

Not accounting for such heterogeneity across rich and poor is even more problematic if we use relationships documented in countries where good data is available (usually rich countries) to project estimates for countries where data is unavailable (usually poor countries). Unless we have accurate data on deaths for all countries of the world, our only option for thinking about the poorest and hottest places is to interpolate. And to do this well, we need, at minimum, some sense of whether a country's average income or air-conditioning penetration or other forms of adaptation affect the temperature–mortality relationship.

This may be especially important, given the possibility that we may mistake the relatively small effect of heat documented in richer places using more readily available data as an indication that heat may not be a big deal in the grand scheme of threats to health.

New Approaches, Sobering Statistics

Motivated in part by the prospect of climate change, a group of economists sought to address some of these issues in the late 2000s. Among them were Olivier Deschênes and Michael Greenstone, who together compiled a database of 72 million death certificates for the United States, spanning the period 1968 to 2002.[19]

Instead of limiting the analysis to death certificates that list extreme temperature as a contributing factor (the *classification approach*), they included deaths of all causes. So every person who died of heart attack, cancer, even automobile crashes are included. And instead of focusing solely on salient events—like the 1995 Chicago heat wave (the heat wave case study approach)—they looked at the effect of all temperature conditions, hot and cold. They included the many days each year when temperatures are warmer—say in the 80s and 90s (°F) (roughly in the upper 20s and lower 30s [°C])—but nothing resembling a major heat wave.

Such efforts have enabled a more comprehensive assessment of temperature's hidden consequences for human health, and sparked a wave of research that leads us to a vastly different understanding of the magnitude of the problem.

These methodological innovations, including their careful utilization of the natural experiment, constitute an important theme of this book. They are part of a broader evolution in economics also often referred to as the *credibility revolution*, which eventually resulted in the 2021 Nobel Prize in Economics being awarded to Josh Angrist, David Card, and Guido Imbens, leaders in the field who ushered in a change in the way economists use data to assess theories.

In short, they looked at how changes in the number of hot (or cold) days in a given county and year affected the number of deaths (specifically, the mortality rate per 100,000 people) in that county-year, relative to the county's average. So if Wayne County, Michigan, experienced 1,000 deaths per 100,000 people on average, but in years with more hot days—holding everything else fixed—experienced 1,010 deaths per 100,000 people, they would infer that the hotter days led to a 1 percent increase in mortality. In another year that was even hotter, if the total death rate was 1,040 deaths per 100,000, they would infer that heat led to a 4 percent increase in mortality, and so on.

It is important to note that the analysis compares changes in mortality within a given county to changes in temperature within the same county.[20] This means they are able to control for factors like average income, access to healthcare, or prevalence of smoking, which vary by location and which, given patterns of development and racial composition in the United States, may be strongly correlated with average temperature in the cross section that may otherwise lead to bias (we will return to this *omitted variable bias* in the next chapter). Because they look at how temperature over the course of an entire year affects overall mortality that year, they can account for the possibility of forward displacement, which the literature suggests typically tapers off around thirty days. They were also far less prescriptive in determining what constitutes a relevant temperature event. Instead of imposing an ad hoc definition of a heat wave (or cold spell), they simply counted the number of days each year in one of twenty temperature bins, ranging from days below 0°F (−17.8°C) to days above 100°F (37.8°C), including separate variables for the number of days in the 70s, 80s, and 90s (°F). This meant that, if even moderate heat (or cold) events had the effect of increasing mortality, they would be able to pick it up in the analysis.

They also accounted more explicitly than any previous study for the fact that individuals would, if they had the means to do so, engage in adaptive behaviors to try to minimize the health risks of heat—notably, by increasing their home energy usage to run a fan or air conditioner.

The implications of this approach are sobering. By these estimates, an additional day with average temperatures in the 90°F (32.2°C) range—which is likely one with highs in the upper 90s or 100s (°F), given the inclusion of nighttime lows in the calculation of daily mean temperature—increases *annual* mortality by 0.1 percent. As a rough approximation, which requires many important assumptions, one might think of this as an increase of around 35 percent on the baseline *daily* mortality rate, relative to a day in more optimal temperature ranges.[21] Another way to think about this number is to note that, if every American were to experience a single additional day with mean temperatures above 90°F, and one fewer day in the 60s (°F), this could lead to approximately 3,000 additional deaths that year.

Let us unpack that number. The baseline mortality rate in the United States is roughly 1,000 per 100,000. That is, one in every 100 people pass away each year on average. So, in a country with 300 million people, that's roughly 3 million deaths each year.

In a year with one more really hot day (mean temperature above 90°F), that rate goes up by approximately 0.1 percent, from 1,000 per 100,000 to 1,001 per 100,000. Multiplying this out across the US population gets us to 3,000.

Once again, this is a very rough approximation because technically this number represents the additional mortality burden associated with taking one mild day (say, temperatures in the 60s [°F]) and turning it into a very hot one (with highs in the upper 90s and 100s), holding everything else constant.[22] But it points to the possibility that hotter temperature may exert a more pervasive impact on human health than official statistics of heat-related deaths might suggest.

It is also worth noting the effects on the other end of the temperature spectrum. Researchers, including Deschenes and Greenstone, have found using similar methods that colder temperatures significantly increase mortality as well. In fact, in many countries including the United States, the United Kingdom, Germany, and Japan, colder temperature appears to kill many more people than heat on average. While there are still unresolved puzzles regarding the mechanisms at play, this pattern is consistent with the links between colder temperature and human physiology, as well as an increase in influenza and other infectious diseases that become more prevalent during periods of extreme cold, when people are more likely to be indoors. Because the cumulative magnitude of these impacts is significant, they will be important in thinking about the *net* effect of climate change, particularly in terms of inequality between a country's (or the world's) cooler and warmer (often richer and poorer) locales.

As we will see later, the profile of asymmetry in the effects of hot and cold temperature work out in such a way that, for many parts of the world, the net effect is to *increase* mortality, including in the United States. But there are countries like Sweden, Norway, and Canada where, even despite the larger marginal effect of a hot day, the reduction in deadly extreme cold days is expected to be so large that climate change may represent a net *reduction* in mortality. We will come back to the regional and global inequality arising from this asymmetry in part III.

These caveats notwithstanding, the thought experiment is instructive. One more very hot day routinely killing at least 3,000 people in the United States each year is on par with the number of deaths from the terrorist attacks of September 11, which caused 2,977 fatalities. It is several times the CDC figure from before (700 to 800).

Importantly, this simple calculation does not include the increased mortality due to less extreme heat—that is, days with mean temperature in the 80–85°F (26.6–29.4°C), or 85–90°F (29.4–32.2°C) range. In fact, studies increasingly show that the cumulative consequences of these milder instances of heat may be just as deadly, if not more so.

A recent study by a trio of economists looked at this question using data from tens of millions of US senior citizens, based on administrative data from government health insurance records. Using similar techniques, they found that, while days above 95°F led to the sharpest increase in mortality, days in the 80s and lower 90s also increased deaths significantly. A day between 80°F and 85°F (26.6°C to 29.4°C) increased the mortality rate by 0.25 in 100,000 seniors age 65 and above. A day between 85°F and 90°F (29.4°C to 32.2°C) increased it by 0.5 in 100,000; a day between 90°F and 95°F (32.2°C to 35°C), by 1 per 100,000. Because there are approximately 25, 5, and 3 times as many days in these "less intense" hot days, respectively, than the hottest category (> 95°F; > 35°C), the vast majority of the additional mortality from hotter temperature appears to be driven by such unremarkable but nevertheless dangerous heat events. In fact, their estimates suggested that more than two-thirds of the increased elderly mortality due to hotter temperature comes from these milder heat events, few of which would be classified as heat waves worth covering in the news.[23]

These and other estimates suggest that hotter temperature may have been responsible for the deaths of close to 10,000 senior citizens in the United States *each year*,[24] far more than the 700 to 800 deaths per year in the entire US population listed in official records. Looking forward, these estimates suggest that climate change could result in a sharp increase in elderly mortality.

Again, the piece of the climate change puzzle that gets less attention is the marginal impact of these less extreme but more numerous events affecting more people in total, and the fact that, even in one of the richest and most highly air-conditioned countries on the planet, relatively unremarkable heat events may routinely kill tens of thousands each year.

There are good reasons to suspect that the magnitude of heat-related mortality may be much higher in developing countries. We will examine some of the hypothesized reasons more closely in part III. For now, some headline statistics are notable. One study found, using similar data and methods to the Deschenes and Greenstone study above, that a 90°F day increases annual mortality by 1 percent in South and Southeast Asian countries relative to days with more optimal temperatures.[25] This is approximately ten times the effect in the United States. In a place like India, where the baseline death rate is roughly 900 per 100,000 and which has a population of 1.4 billion, this means that if everyone in the country were to be exposed to a single day with average daily temperatures above 90°F (32.2°C), the result would be greater than nine per 100,000 deaths, which comes out to 126,000 extra deaths. Most likely a tiny fraction of these deaths would be categorized as heat-related, and they would be diffused across the entire country, meaning that we would only understand the true toll in retrospect.

Understanding the causes of such higher vulnerability to heat and other climate hazards is of vital importance, especially given the intimate relationship between economic inequality and climate change. Better diagnosing the specific causes of sources of vulnerability will also be critical in doing something about it.

Bringing it back to the United States, according to recent climate projections, business-as-usual emissions could lead the number of days with mean temperature above 90°F in the average county in the United States to increase from about one per year (pre-2010) to about forty-three per year (2070–2099).

Linearly projecting these numbers out to 2070 and beyond involves some heroic assumptions, especially since people will likely adapt whenever they can, and no one knows what technology or the economy will look like by then. We will discuss in greater detail how researchers attempt to think carefully about such changes and how to incorporate them into climate damage projections. For now, the point is that available estimates suggest that the health impacts of hotter temperature alone may be much larger than we used to think. In case you don't have a calculator handy: $3,000 \times 43 = 129,000$.

The Global Mortality Burden of Hotter Temperature

As of this writing, researchers have compiled the data necessary to provide an updated, if still incomplete, estimate of the global mortality burden associated with extreme temperature: one that considers the effects of both the very extreme and lesser extreme heat events, and incorporates the possibility that being richer (and more accustomed to heat) can help people adapt. For instance, the Climate Impact Lab, a research consortium involving the University of California, Berkeley, the University of Chicago, and the Rhodium Group, have estimated detailed heat-mortality response functions for 24,378 regions roughly equivalent to US counties in forty countries, generated from comprehensive mortality data spanning several decades, representing approximately 55 percent of the world's population.[26]

Importantly, thanks to this research we now know much more about how the temperature–mortality relationship varies, both by average income and average climate. This means we can more reliably project not only the current health impacts of hotter *weather* in the short run, but also the potential mortality consequences of a hotter *climate* in the long run, accounting for many of the adaptive responses that humans will no doubt engage in over time. It also means that, with some important provisos, we can make better-educated guesses about the effects of heat on health in many poorer and hotter parts of the world like sub-Saharan Africa, where lack of high-quality data has hindered such estimates in the past despite potentially outsize climate impacts. Because the estimates rely less on ad hoc assumptions about adaptation but allow the data to tell us just how much adaptation we might expect as incomes rise and our experience with extreme temperature grows, projections about the distant future can be made with more confidence and transparency than previously possible.

According to the Climate Impact Lab's estimates, climate change may cause an increase in global mortality rates of seventy-three deaths per 100,000 per year by the end of the century. This is despite (indeed, explicitly accounting for) the fact that by 2100, average incomes will be significantly higher. It also accounts for the fact that, even holding income constant, people will likely adapt to warmer temperatures over time. In fact, without these adjustments, the implied increase in mortality due to climate change would be 221 deaths per 100,000, which is nearly three times the estimate without accounting for adaptation and income growth.

Let us put the lower figure of seventy-three out of 100,000 number in perspective. This is a massive number. It is roughly two-thirds of the US COVID-19 fatality rate in 2020. It is *six times larger* than the current fatality rate from automobile accidents in the United States. It is also orders of magnitude larger than the 0.33 per 100,000 increase estimated using one of the existing integrated assessment models (i.e., the Framework for Uncertainty, Negotiation, and Distribution [FUND]), which as we will discuss in part IV are important tools that policymakers use in assessing the costs and benefits of reducing emissions.

It is worth underscoring the significance of adaptation. Without accounting for adaptation, the projected increase in mortality due to hotter temperatures is 221 deaths per 100,000. For a sense of magnitude, this is on par with all global deaths from cardiovascular diseases today, which is the world's leading driver of mortality by far.

What accounts for the gap? We will revisit the issue of adaptation and resilience in later chapters. It turns out that they may be just as important as the actions we take to reduce emissions, and that what constitutes *adaptation* may not always be what you expect.

This study is both a scientific feat and a real wake-up call. It represents a culmination of decades of refinement of empirical tools and data collection, motivated in part by the objective of more accurately understanding the true social cost of hotter temperatures. The numbers suggest that most of our official records when it comes to the health consequences of heat may vastly understate the seriousness of the problem—both in terms of the overall death toll and the profile of health inequality.

Two important caveats are worth noting here. The first is that these projections must by necessity make some assumptions about the pace of future technological change and how this will affect the baseline assumptions embedded in these and other projections, especially regarding how societies will adapt. The second is the near-complete omission of the African continent from the underlying data that informs these estimates, mainly due to lack of high-quality data, though the researchers provide extrapolated estimates of potential impacts in Africa based on observed temperature–mortality relationships of similarly hot and poor parts of Asia and the Americas. There may be limits to the validity of such an extrapolation if other factors, like drought or conflict or disease burden, mediate the relationship between temperature and mortality differently across continents.

It is unclear in which direction these two factors may bias the estimates. For now, it is worth emphasizing just how much of a scientific advance this study represents in our ability to empirically estimate future climate damages, and just how many people may die due to heat in the coming decades.

Another way to think about the magnitude of these findings is to return to our quiz and place this mortality burden in the context of other leading causes of death. The last question asked us to consider how many people, in any given year, would still be alive were it not for hotter temperatures.

As mentioned before, if one stops to think about it, it is not the most straightforward hypothetical because, unlike AIDS or cancer or malaria, it is difficult to imagine a world in which "heat" is eradicated. It would be akin to asking what would happen if everything else about the world remained the same, except that everyone experienced year-round temperatures of around 75°F (23.8°C) or below (or had enough air-conditioning that heat no longer impacted mortality). For these and other reasons—including the fact that it is difficult to imagine all other patterns of human activity relevant to health to stay constant as one changes experienced temperature so drastically—the Climate Impact Lab researchers stop short of this sort of present-day hypothetical.

By my own back-of-the-envelope calculations using Climate Impact Lab estimates, somewhere in the vicinity of 500,000 to 1,000,000 fewer people per year may have died prematurely, were it not for hotter temperatures during their study period.[27] This is on the same order of magnitude as other recent studies using similar methods to estimate the total number of deaths attributable to "non-optimal" temperatures, which one study by Qi Zhao and coauthors estimates to be over 5 million per year in total, with approximately 480,000 coming from heat, the rest from colder temperatures.[28] It is somewhat less than estimates of the number of people who die due to traffic accidents each year, and on par with AIDS or malaria. The CDC reports that, globally, 680,000 people died in 2020 due to AIDS; another 627,000 people due to malaria.[29]

Most of these deaths are occurring not during extreme, headline-making heat waves, but during more mundane episodes of hotter temperature. And the vast majority are not classified in official records as being caused by heat, but rather tagged as deaths due to heart disease or other respiratory or endocrine diseases.

In case you thought you could avoid those pesky two-handed provisos, it is important to note additional limitations of this thought experiment. First, it is entirely possible that other diseases exhibit significant undercounting in the official records. For instance, many individuals who have chronic ailments such as diabetes are not aware that they do, or never receive a formal diagnosis. The CDC estimates that whereas 37 million adults aged 18 years or older in the United States had diabetes in 2019, 8.5 million adults who met laboratory criteria for diabetes were not aware of or did not report having diabetes (undiagnosed diabetes). This number represents 3.4 percent of all US adults and 23 percent of all US adults with diabetes.[30] But here we are talking about conditions that can be reliably diagnosed using only the information provided by the body itself. We are not inferring the conditions under which a particular death or injury arose.

Second, it's worth repeating the slightly contrived nature of the hypothetical. The number above is in comparison to a hypothetical world in which we have eliminated heat-induced mortality, and using estimates of the deadliness of hot days relative to "optimal" temperature days. What this corresponds to requires a bit of imagination. Perhaps it is a world where everyone lives in a climate like Berkeley, California, or the Canary Islands in Spain where the temperature rarely if ever reaches extremes in either direction but not everyone has regular access to cooling and heating technologies, as they retain all other features of their economic, social, and physical environment.

The point of the thought experiment is not to generate a replacement statistic for official use but to get us to start thinking about the less visible damages associated with a hotter climate. Evidence from other domains, like workplace safety, suggest that there may be a broader pattern at play. For instance, in my own research, my colleagues and I find a similar divergence between official heat illnesses and workplace accidents caused by hotter temperature more broadly. Official records kept by the California Division of Occupational Health and Safety (Cal/OSHA) count 672 heat illnesses over the period 2007–2019, roughly 52 per year in California.[31] According to work by my colleagues Nora Pankratz and Patrick Behrer and I using a comprehensive database of worker's compensation claims from the state—which includes all types of injuries not limited to heat illnesses, but also including bruises, cuts, falls, and more—a day above 80°F leads injuries to increase by 5 to 15 percent compared to days with temperatures in the 50s and 60s (°F). While reasonable minds can disagree on what constitutes the correct baseline temperature, as with the heat-mortality calculations above, it seems plausible according to our estimates that upward of 15,000 workplace injuries in California may be caused by hotter temperature each year, in that they would not have occurred if the work had taken place in a cooler thermal environment. This is a figure many times greater than that of the official count of heat illnesses in California during the same time period.

So how did you do on the quiz at the beginning of the chapter? According to recent estimates, the direct mortality associated with a hot climate may already be on par with AIDS, and rising quickly.

One way to think about the threat of climate change is that it may lead hotter temperatures to overtake all accidents, and perhaps even all cancers, as a leading cause of death by the end of the century if emissions aren't reduced significantly. In fact, without any adaptation (which seems unlikely but can be informative as a hypothetical), and with business-as-usual emissions, it is not implausible that climate change could lead to an increase in mortality similar to heart disease today, which is the leading cause of death by far.[32]

While it is difficult to predict what will happen at temperatures outside of observed experience, the available evidence indicates that a nontrivial share of these deaths are likely to come from what one might call the more mundane heat events—from a steadily growing preponderance of days with temperatures hot enough to be deadly for some, though not hot enough to qualify as extreme heat waves.

Stories of unlivable cities and mass deaths due to mega heat domes can generate eye-catching headlines. There is even a movement to start naming heat waves the way we name tropical storms.[33] Attaching psychological salience to such one-off events may be important as tools for rattling us out of our complacency. But as the evidence presented in this chapter suggests, it is important to remember how the less extreme but more numerous heat events may affect the magnitude of overall human suffering realized.

6

Temperature and the Wealth of Nations

LET US RETURN to the experiences of LeBron James and Larry Bird, as well as of Sofia and her classmates in New York City's public schools. There we witnessed some of the challenges of assigning causality to temperature in settings where so many other factors may also be at play: where all-else-equal counterfactual comparisons are not readily available. We also saw how the mind, when presented only with anecdata and undisciplined by the scientific method, can be quick to project biases and stereotypes in interpreting the social consequences of natural phenomena—including the implications of hotter temperature.

In this chapter, we return to a fundamental question: What is the relationship between temperature and human performance? In the process, we will delve deeper into the scientific innovations that help shed new light on this old question, and how these insights may inform our understanding of the economic consequences of climate change.

While global warming pushes us to consider how this question factors prospectively, they also matter retrospectively—for our understanding of how economic development got us to where we are now and about the stories we tell ourselves regarding the role of climate in shaping our individual and collective destinies.

Here are some statistics worth noting. Globally, 1.3 billion individuals (roughly 16 percent of the world's population) work in agriculture or construction, two industries where most of the work occurs outdoors.[1]

This is four times the number of people living in the United States. As late as the 1980s, that number was nearly double what it is now.

Even in the United States, over 70 percent of the roughly 100 million workers without a bachelor's degree report routine exposure to harsh environmental conditions at work, including extreme temperatures in many indoor jobs such as in manufacturing and warehousing.[2] Worldwide, fewer than 30 percent of households are estimated to have access to air-conditioning. Even within rich European countries such as Sweden or France, residential air-conditioning penetration is less than 50 percent.

The amount of heat that each of us experiences each year varies enormously. The average American experiences 30 days above 90°F (32.2°C); the average German, 4 such days; the average Canadian or Brit, fewer than 2. Compare that to over 100 such days per year for someone living in Accra, Ghana; 150 for the average person living in Mumbai, India, or Bangkok, Thailand; and over 200 for residents of Abu Dhabi, in the United Arab Emirates. Variation within countries can be just as large. While San Francisco averages one or two such days per year, Houston experiences over 100.

The implication is that whether heat has some effect or no effect on human performance (or none indoors but some outdoors) may be important in understanding climate's effects on economic productivity in the long run. Even if there is only a very small effect on the margin, if heat happens frequently enough, the damages could add up for some groups quickly.

There is also the following statistic. Countries that are 1°C (1.6°F) hotter on average have average per capita incomes that are approximately 8 percent lower. It's worth repeating this. Countries that are 1°C hotter on average have per capita incomes that are approximately *8 percent lower on average*.

One can see this relationship visually in figure 6.1. Moving from top to bottom, countries are poorer. Moving from left to right, countries are hotter. So much so that a country that has a climate that is 6°C hotter on average is associated with living standards that are roughly half of its counterparts to the cooler north.

This relationship holds across a wide range of outcomes of interest. Instead of income per capita, we could have plotted life expectancy or

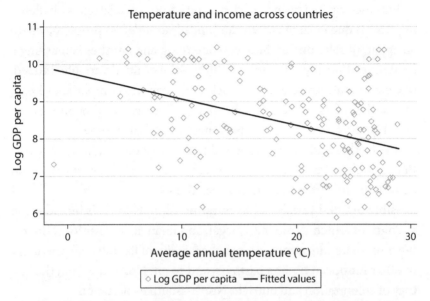

FIGURE 6.1. The relationship between population-weighted average annual temperature and log GDP per capita by country in 2000. *Source*: Author's calculations using data from Melissa Dell, Benjamin F. Jones, and Benjamin A. Olken, "Temperature Shocks and Economic Growth: Evidence from the Last Half Century," *American Economic Journal: Macroeconomics* 4, no. 3 (January 2012): 66–95.

standardized performance on internationally harmonized exams on the *y*-axis. The negative correlation would have been equally striking. Indeed, students in countries that are 1°C hotter on average perform 8 percent of a standard deviation worse on the Programme for International Standardized Achievement exams, which means that a 3°C hotter average climate is associated with nearly a year less of formal schooling for a given age or grade.[3]

This correlation between temperature and indicators of economic advantage persists even within countries, albeit with a shallower slope. For instance, across the United States, Mexico, and much of South America, municipalities that are hotter are poorer on average—by roughly 1.3 percent per degree Celsius hotter average temperature.[4]

Whether temperature plays a causal role in such relationships may be crucial in understanding the economic consequences of climate change.

But once again, teasing apart causality from correlation is challenging. This is due in large part to the *omitted variable bias* problem, where an apparent relationship between one thing and another is masking a hidden relationship with a third thing—the omitted variable. Indeed, one could make the claim that such cross-country comparisons beget the mother of all omitted variable bias problems, given the number of possibly correlated third variables affecting global income.

The Netherlands is far richer and colder than poorer, hotter Nigeria. But its cooler climate is one of a great many other differences that could be driving the differences in income. The list of other differences—of possible omitted variables—is mind-numbingly long. It includes the strength of Dutch democratic, legal, and financial institutions, the robustness of its education systems, the fertility of its soils, the proximity to other European trading partners, and so on. Not to mention the history of colonization and outright North–South exploitation.

Moreover, exceptions to the rule are not difficult to find, and seemingly belie grand theories of climate determinism. For instance, you might notice that there are a handful of countries in the upper right quadrant, representing those that are rich despite being in very hot climates. Who are these countries? They are the Singapores and the Qatars and the Saudi Arabias of the world. Income per capita in Singapore today is over $72,000 ($116,000 purchasing power parity [PPP] adjusted), which is higher than that of the United States. Clearly, a hotter climate does not necessarily doom a people from achieving economic prosperity.

You may also notice a few countries that are cold and very poor in terms of material standards of living. These include Mongolia, Uzbekistan, and North Korea. Indeed, we need not look any further than the Korean peninsula for evidence that seemingly flies in the face of the theory that temperature could be the cause of worse economic outcomes. Even though North and South Korea are separated by what is arguably a randomly drawn line across a similarly temperate peninsula the size of the United Kingdom, their economic fortunes could not be more divergent. The average South Korean earns over $34,000 per year ($47,000 PPP adjusted), over fifteen times the average North Korean.

So, we might want to exercise caution before drawing any conclusions from this correlation. All the more so when considering the extrapolation of such relationships to the context of future climate change. Indeed, getting to the bottom of this relationship, if it is possible at all, will likely require more than simple eyeballing, anecdotes, or even a very large amount of data.

Yet, students of history know that this has not prevented many from making the armchair observation that a hotter climate may pose impediments to human development. The ancient Greeks, including Aristotle, often expounded such ideas, as did many Enlightenment thinkers, including Montesquieu, who in 1748 proclaimed that an "excess of heat" made men "slothful and dispirited."[5]

In more recent centuries, such arguments were wielded alongside theories of "dominant races" and other justifications for colonialism. In his 1915 treatise entitled "Civilization and Climate," political scientist Samuel Huntington noted that "we know that the denizens of the torrid zone are slow and backward, and we almost universally agree that this is connected with the damp, steady heat." He goes on to conclude that, "if our diagnosis is correct, we may at last hope to be free from the withering blight which has overtaken every race from which the stimulus of a good climate has been removed."[6]

Perhaps this is part of the reason why, for much of the twentieth century, many social scientists were hesitant to test related (less racially charged) hypotheses regarding the relationships between climate and economic development, despite their potential relevance today.

The specter of climate change arguably makes better understanding this relationship urgent, cultural baggage notwithstanding. However, to do so we need to first get a handle on this question of causality. Does temperature *cause* human performance to decline? How does one separate out such an effect from other correlated factors?

Evidence from the Laboratory

Evidence from laboratory experiments provides some helpful clues.

Back in the early 1950s, the British Navy asked a team of researchers to "enquire into the effects of heat on human beings." A major motivation appears to have been to manage costs. At the time, many ships in Her Majesty's Fleet were stationed in colonial and former colonial outposts such as Egypt, India, and Nigeria. And while the movement to decolonize was gaining steam, the British Navy maintained a sizable presence throughout the tropics until the 1960s and beyond. The British government wanted to figure out which ship's compartments should have priority access to expensive air-conditioning equipment.

A team of scientists led by Dr. N. H. Mackworth undertook a series of laboratory experiments by enlisting volunteers from within the navy's ranks. The subjects ranged from ages 18 to 34, and had varying levels of previous military experience, some having served multiple tours in such far-flung corners of the British Empire as "Ceylon" and "The Middle East and Persia." All subjects were men.

In one study, a group of eleven naval officers were placed in a room and asked to perform repetitive tasks—including interpreting Morse code. On arrival at the lab, these men were each fitted with a rectal thermometer, which is probably as uncomfortable as it sounds. Thermometers snugly in place, the officers were asked to interpret as many Morse code messages as they could, as accurately as they could, under varying temperatures. They repeated this exercise over the course of five days, for up to three hours at a time. Sometimes the ambient temperature was set at 75°F (23.8°C), other times it was as high as 105°F (45°C). The officers' activity was measured and meticulously documented (see figure 6.2). The various officer groupings were organized so that different groups experienced hotter temperatures on different days, in part to separate out any confounding effects due to improvement in coding skill or deterioration in officer effort over time.

The results were striking. Subjects made substantially more mistakes under hotter conditions, even controlling for individual differences in coding ability. Whereas the average officer made roughly 11 to 12 mistakes per hour at temperatures between 85°F (29.4°C) and 90°F (32.2°C), the error rate went up to 17 per hour at 100°F (37.8°C) and skyrocketed to 95 per hour at 105°F (40.5°C). Mackworth and his colleagues also noted that the error rate appeared to rise as the hours wore on, with more errors occurring in the second hour compared to the first, and more still during the third compared to the second, regardless of the temperature. They took this evidence as indicative of fatigue.

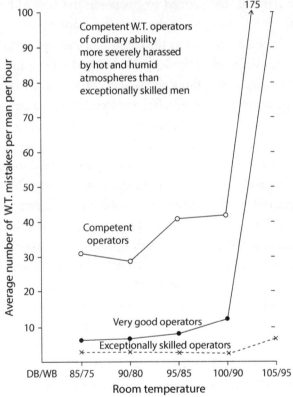

FIGURE 6.2. The relationship between room temperature and the performance of British naval officers. The graph below the photo shows the average number of mistakes made by wireless telegraph operators as a function of room temperature, plotting separate lines for "competent," "very good," and "exceptionally skilled" operators. *Source*: Adapted from Norman H. Mackworth, "Effects of Heat on Wireless Operators," *British Journal of Industrial Medicine* 3, no. 3 (1946): 143.

Interestingly, and despite the discomfort endured, no significant relationship was found between rectal temperature and officer performance, in part because baseline core body temperature appeared to vary substantially across individuals, and because the bodies of these otherwise physically fit individuals appeared to maintain relatively stable internal temperatures despite external heat, at least up until the heat reached very high levels. Thankfully, the researchers did not push their subjects too much further beyond the physical safety zone.

Let us set aside for a moment any potential problems of basic experimental design.[7] The study illustrates the value of controlled experimentation. Namely, the possibility of getting us closer to *all else equal* comparisons, or the counterfactual that proved so elusive in the case of LeBron and Magic in the NBA examples in the previous chapter.

Indeed, there is a reason why the controlled and replicable laboratory experiment is a hallmark of the scientific method. Create an artificial environment in which everything else is held constant and manipulate the potential causal factor of interest. If the outcome that you care about (e.g., coding accuracy) varies with changes in this factor (e.g., temperature), then one can have much greater confidence in the causal significance of it.

Since Mackworth and the British naval experiments, dozens of laboratory experiments have documented similar relationships between temperature and task productivity using similar methods. Across a range of outcomes, including accuracy at guiding a steering wheel or engaging in routine manual labor tasks such as lifting boxes, these laboratory studies document an inverted U-shaped relationship between temperature and human performance. The general pattern is one wherein standardized performance declines by one to two percentage points (in some cases more) per degree Celsius above room temperature, typically between 20°C and 24°C (68°F and 75°F).[8] One recent metareview of such studies suggests that performance at 30°C (86°F) is roughly 9 percent lower than at 22°C (72°F).[9] While cold was not a concern for Mackworth and his colleagues, subsequent studies also suggest that colder temperatures can reduce performance as well, though to a lesser extent.

These findings are consistent with subsequent medical studies, which find that many systems of the human body are sensitive to temperature.

Hotter temperature can lead to increased blood viscosity and blood cholesterol levels and can reduce cardiovascular efficiency. When body temperatures rise, more blood flows to the skin and the body sweats to dissipate heat in order to maintain thermoregulatory balance. A fever occurs when this balance tips and body temperatures rise above what is optimal for the body's many systems to function properly.[10]

The brain may be especially sensitive to temperature stress. Not unlike a CPU that heats up with use, the human brain creates heat as its neurons fire. In fact, while the brain makes up only 2 percent of body mass, brain tissue is responsible for approximately 20 percent of the heat released by the human body.[11] Normally, this heat exhaust is carried to the external environment via the bloodstream and the skin or lungs. When the ambient temperature rises, this transfer of heat out of the body into the environment slows, as there is less of a gradient between the body and the surrounding air.[12]

While there is still disagreement regarding the specific pathways through which temperature affects brain function, and whether certain cognitive processes are more sensitive than others (including, as we will discuss in chapter 7, the processes that govern our emotional impulses and the regulation thereof), a growing number of laboratory studies confirm the general idea that temperature can affect human performance, cognitive and physical.

But a closer look at the British Navy studies reveals some important limitations of controlled experiments as a category, especially if what one is interested in are the potential human consequences of living in a hotter climate. Let us focus on three.

First, it is unclear to what extent laboratory settings replicate the incentives faced by people in the real world. How much discomfort and fatigue do we expect subjects to put up with when all that is at stake is the potential disappointment of the scientist studying their behavior? Do we expect similar reductions in performance when the stakes are higher—say during important exams or athletics matches or job interviews?

If we believe human beings respond to incentives, even nonmonetary ones such as social pressure or personal pride, we may want to find better matches between the incentives facing subjects in a lab and the incentives that prevail in the real-world contexts we care about most. But to run experiments in such high-stakes settings may in some cases be unethical, precisely because we would be messing with people's lives. The horrors of the Tuskegee experiments of the 1930s, when treatments known to be effective for syphilis were withheld from Black subjects, are painful reminders of what can happen when such ethical boundaries are not respected.[13] In other cases, such real-world experiments would simply be very costly to administer.

Second, it is difficult to infer whether such findings can be applied to other demographic groups. All subjects in the Mackworth study were young, physically fit males: white males, and officers in the British Navy at that. How much does their heat tolerance reflect that of infants or the elderly? Or men of other races and ethnicities? Or of women? Those who have quarreled with a spouse or a child or a grandparent over control of the bedroom thermostat may have some intuition that such tolerances may vary.

Third, there is a complete lack of scope for adaptation. By design, such laboratory experiments *constrain adaptation* to be essentially zero. You're stuck in a hot room. You've been given a task to complete. Your options are highly limited by construction.

How much does this conform with the real world? If you are a farmer tending a field, or a construction worker helping to build a highway, can you plan your day to avoid the heat of the sun? If you are the manager of a small manufacturing facility, or a freelance writer working from home, would you have an incentive to cool the shop floor or the home office?

Relatedly, it is unclear how much the effect of temperature during a relatively short period of time—three hours at a time, in the case of the Mackworth study—is reflective of what happens when someone is faced with repeated or periodic exposure over a longer period, say over multiple shifts in a warehouse of multiple years in an non-air-conditioned school. On the one hand, the scope for adaptation may increase as the window of exposure lengthens. On the other hand, the damaging effects may be compounding, and the cost of adaptation may also add up in themselves.

Notice how the limitations associated with economic stakes and demographic representativeness may interact in important ways. How much of an effect on productivity do we expect for individuals in resource-constrained settings, where access to adaptation technologies such as fans or heat pumps or air-conditioning may be limited, compared to those without such constraints? And to what extent do we think other factors associated with poverty—prevalence of chronic diseases, baseline crime rates in the neighborhood, access to banking or social insurance—affect the capacity to make adaptation investments, or the propensity to engage in adaptive behaviors? For instance, are people living in high-poverty, high-crime neighborhoods less likely to go outside or open their windows on hotter days?

We can infer from laboratory studies that temperature can, in highly controlled circumstances with limited stakes, affect human performance. We know this much with a high degree of confidence. The strength of such lab experiments is in establishing causality in a narrow sense (strong internal validity). We know that temperature was most likely the only thing that varied, because the experiments were set up that way. But such laboratory experiments have limitations in what they can tell us about the many real-world complications that matter in thinking about the effect of a hotter climate on human well-being (limited external validity). For this, we need data from the real world, and a disciplined way of teasing apart cause and effect without the benefit of a randomized experiment.

Evidence from the Field

Let us return once more to Sofia of New York City. Or, more precisely, the data from which statistically representative Sofia was constructed. Suppose we could compare Sofia's performance, across all her exams, with the temperature on each of those exams. Would this get us closer to the truth? What if we could do this not just for one student, but all the students who took Regents exams in New York City that year and in other years, and for all subjects ever tested: might there be a way to harness the power of the scientific method without conducting an actual experiment, to make it as if nature had run a series of experiments with real-world consequences?

Thanks to the meticulous recordkeeping of New York City public schools, the test scores are there, all 4.5 million of them. Though temperature readings from test sites themselves may not be available, one can reconstruct the ambient temperature during the time of each exam in the vicinity of each of the 800-plus schools in the district thanks to daily weather records and satellite imaging. Because the test data includes information on where the exams were taken, when they were taken, and by whom, they can be linked up to estimates of temperature leading up to and during each exam. Recall that, because of the harmonized and rigid administration of these exams, we can be reasonably confident that students were not self-selecting into hotter or colder exam times, which, while possibly adaptive for the students, would have been problematic from a research perspective.

Finally, thanks to a revolution in the way social scientists think about natural experiments, as well as rapid improvements in computing power, we are increasingly able to design quasi-experiments and run statistical analyses that more credibly isolate cause from mere correlation.[14]

The underlying intuition is simple. By comparing differences in performance for a given student across multiple tests and doing so controlling for that student's average ability and the average difficulty of each subject being taken, one might be able to tease apart the influence of hotter temperature on performance as opposed to other potentially correlated factors.[15]

The important feature here is that we are comparing the same student's performance over time, as opposed to comparing students in different places during a snapshot in time. The latter would be problematic if, for instance, students in more densely urban and less green parts of the city tended to be lower-income, or different sets of students were taking the tests on different years. We might mistakenly attribute a correlation between hotter average temperatures (due to urban heat island effects and lack of tree cover) with lower performance (which tends to be correlated with income) as a causal effect of hotter temperature on student performance. This is the Netherlands–Nigeria omitted variable problem all over again, just on a smaller scale.

By looking at differences in performance "within" a given student, we are relying on the variation in weather on any given day, which is not subject to such omitted variable problems, and certainly not within the control of any human being. As long as students are not choosing when and where to take their exams (which is reasonably well assured by their stringent administration), we can be pretty confident that the temperature that each of them experienced during any given exam is more or less randomly assigned.

In other words, where Mackworth and colleagues deliberately manipulated temperature in a laboratory setting, here we can exploit natural variation in temperature occasioned by Mother Nature. And by repeating such "within-subject" comparisons over hundreds of thousands of students taking similar exams, we might more credibly pinpoint the consequences of temperature for human performance in settings where the stakes would be economically meaningful. Whereas looking at the case of one individual student may not allow us to separate real differences due to temperature from random differences due to chance, comparing differences across many thousands of students gives us a better shot at separating the signal—if there is one—from the noise.

And by doing so in a setting where the stakes are economically meaningful, we might have better confidence in the applicability of our findings to questions about the role of temperature on human performance in the real world.

You might have noticed a parallel between the causal inferential tools utilized here and those that helped us paint a fuller picture of the mortality burden of nonoptimal temperature from the previous chapter. The underlying intuition is similar. One difference between the two settings is that, in the case of mortality, we compared changes in mortality rates over time in a given geography since it would be impossible to study changes in mortality for a given individual over time.

The results from the above study, which was a chapter of my PhD dissertation, were surprisingly consistent with those from laboratory settings, including the British Navy studies. Students taking their exams under hotter conditions appeared to perform worse. Significantly worse, but perhaps not obviously so.

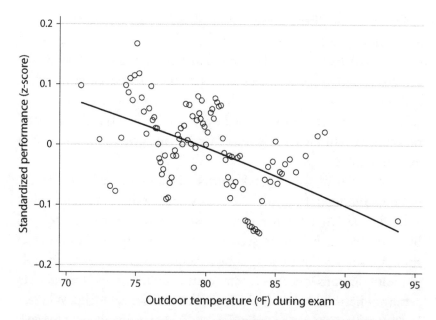

FIGURE 6.3. The relationship between temperature and high school exam performance. The figure shows the relationship between test-day outdoor temperature (°F) and standardized exam performance (standard deviations) for students in New York City public schools, plotting the residuals from a regression that includes school fixed effects and controls for student demographic characteristics including gender, ethnicity, and indicators for poverty. Each circle represents approximately 40,000 test scores. *Source*: R. Jisung Park, "Hot Temperature and High-Stakes Performance," *Journal of Human Resources* 57, no. 2 (2022): 400–434.

Taking an exam when the temperature was 90°F (32.2°C) outside reduced performance by around three to four points, which is small enough to be difficult to pick up with one or even a dozen students' records. But these effects aren't trivial, particularly for the median student, who is already in danger of failing most of these exams. They imply that the average student was roughly 10 percent less likely to pass a given subject than they would have if they had taken the same exam when it was a more comfortable 70°F (21.1°C; see figure 6.3).

These effects had measurable real-life consequences. Because the data allows anonymous linkages between student test results and graduation status, we can see whether students who experience hotter exams were more or less likely to graduate from high school. In fact, students who experience one standard deviation hotter average exam-time temperatures were approximately 4.5 percent less likely to graduate from high school on time. All told, the study estimates that, over the period 1998 to 2011, over 500,000 exams that otherwise would have received passing marks were given failing grades, affecting the on-time graduation prospects of at least 90,000 students. One can only speculate what the ripple effects of these failed exams might have ended up being, but available data from other studies suggest they could have been substantial.

These patterns are broadly consistent with some early studies that found evidence that heat's adverse effects on military performance may extend to settings with higher stakes. One of the earliest such analyses looked at records of 500 helicopter accidents due to pilot errors and found that hotter temperatures led to significantly elevated risk of pilot errors. Considering that the single largest cause of premature mortality among pilots is accidents caused by human error, these findings suggested that the performance deterioration may carry over into settings that are very high stakes indeed.[16]

These and other findings imply that hotter temperatures can significantly affect human performance in the real world across a wide range of economic and social settings. Some of these impacts may be noticeable. Many may not, especially given the myriad other factors that can influence human performance in addition to temperature. The power of the approaches described here is that they give us more confidence that, despite the influence of such a wide range of factors, we can separate the causal signal associated with temperature from the noise.

Could air-conditioning have helped? Reliable school-level air-conditioning data was not available for the New York City study. But evidence from similar settings is suggestive. In a study looking at high school exit (or Bagrut) exams in Israel, economists Victor Lavy, Avraham Ebenstein, and Sefi Roth found that while students taking their exams on more polluted days fared substantially worse, temperature appeared to have little effect despite the hot climate. They noted that, in Israel at least, such exams are mandated to be taken in facilities that have working air-conditioning. In a previously mentioned study where we assessed the effects of hotter temperature, not on the day of the test but during class time leading up to it, my colleagues Josh Goodman, Mike Hurwitz, Jon Smith, and I found that school and home air-conditioning both appear to substantially reduce the negative effects of heat on student learning.

But other recent studies still suggest that, even when the stakes are very high, air-conditioning is not a given during exams, particularly when resource constraints are more likely to bind. Few exams could have more momentous stakes than the Chinese College Entrance exams. The National College Entrance Examination (NCEE), or Gao-kao, is taken by over 9 million students each year—more than the entire population of Austria—across China. Unlike most other countries, where colleges rely not only on standardized tests but a wide range of inputs such as high school grade point average (GPA), extracurriculars, and recommendation letters to determine admissions, the NCEE has long been effectively the sole determinant for college admission.

In a recent study using data from 14 million such exam records, economists Joshua Graff-Zivin, Yingquan Song, Qu Tang, and Peng Zhang found that students taking exams during hotter days performed substantially worse.[17] Taking an exam when temperatures were around 32°C (90°F) resulted in a performance decline of approximately 3.5 percent, which is equivalent to 30 percent of a standard deviation. This decline is notable as it reduces the probability of gaining admission to a first-tier university by approximately 6 percent. Such a finding is of economic significance given the average earnings gap of 40 percent between top-tier university graduates and those from less prestigious universities. It also underscores the critical importance of understanding the long-term impact of environmental factors on economic opportunity.[18]

Temperature and Human Performance Revisited

Though many important questions remain unanswered, these studies move us closer to deciphering the true relationship between climate and economic prosperity noted above.

With the right data and the right setting, the natural experimental method helps us shed new light on the old question of how temperature affects human performance. The tools of causal inference are like the Rosetta Stone that unlocks the meanings hidden in the data: it is a practice in lining up the puzzle pieces in the data so that more confident claims about causality can be made and lets us be more transparent and systematic when claims of causality are not well-founded.

We know now that the physical effects of heat that LeBron and Magic experienced were somewhat predictable, even if the complex interactions that led to those particular sporting outcomes were not. Temperature appears to have a systematic and discernible impact on physical performance, even when the stakes are high for professional athletes as well as for laypeople young and old.

In a recent study, Marshall Burke and his colleagues at Stanford University used data from over 150,000 professional tennis matches over the course of fifteen years. Not just on wins and losses, but point-by-point data on double faults, rally lengths, serve speeds, and so on for a wide spectrum of players—women and men, veterans and rookies, Europeans, Americans, Asians, Africans, South Americans, Australians.[19]

As in the New York City study, the researchers exploit the fact that tennis players have very little control over when matches are scheduled, or who they play against, despite the fact they have a lot on the line. In the case of New York City students, we were looking at changes in an individual student's performance across multiple days of testing. Here, Burke and his colleagues looked at various metrics of athletic performance across multiple matches for a given player.

Because the data is so granular—and crucially, because the same player is observed in the same tournament and playing at a given point in a tournament multiple times—the researchers were able to net out many other situational factors with a high degree of credibility. How accurate are male players' services on average? How much more accurate are they at a major event like Wimbledon? Factors such as these are accounted for, which allows an estimation of how much less accurate Rafael Nadal's serves become on a hot and muggy night, and can be repeated and averaged over the spectrum from Novak Djokovic to Maria Sharapova.

Repeating this experiment across many hundreds of players allows them to answer the question: What is the average effect of temperature on serve accuracy? The effect appears to be substantial. Playing at temperatures of 95°F (35°C) increases the chances of a double fault by nearly 7 percent. Players also appear to tire more easily and attempt to conserve more energy during matches. Average rally lengths decline by 10 percent at 95°F, and the average distance run falls by nearly 10 percent as well.

Hotter temperature appears to affect all players, but some more than others. For reasons that are yet unclear, men appear to be more sensitive to the heat than women. They tend to double fault more often during hotter matches. There is also some evidence that playing in a match in sweltering temperatures significantly increases the chances that a player retires from tennis entirely after that match, but only for men.

Interestingly, the study also provides evidence for what economists call *skill-bias*: the effect of outcomes varying depending on one's average proficiency, favoring better players on average. The most highly skilled tennis players appear to be more resilient to the heat, experiencing a less precipitous decline in performance than those with less experience or demonstrated prowess on the court.

Therefore, there is a case to be made that not all of us are affected by the heat in the same way, and that there is something about performing highly skilled tasks under duress that benefits from prior experience. Nevertheless, hotter temperature appears to affect all of us to some degree.

Once again, the combination of big data and natural experimental method allows more credible claims of causality in settings that are arguably of more economic significance. If tennis players in high stakes tennis tournaments are affected by the heat, it seems plausible that heat makes peak performance more challenging in other settings requiring similar physical exertion too.[20]

Manufacturing and Labor Productivity

A growing number of studies like those mentioned above give us greater insight into the ways heat can affect human performance. They suggest that temperature matters nontrivially, for both cognitive as well as physical tasks, even when the stakes are high and even in settings where cooling is technologically feasible.

The implications for how we understand climate's effect on the economy could be profound. But the mapping from temperature to task productivity to economic output is a messy one, with many complexities we do not yet fully understand. Moreover, the possibility of adaptation means that one cannot draw a simple, straight line from these facts to measures of macroeconomic productivity today or climate's impacts on the economy tomorrow. Nevertheless, emerging findings point to a broader pattern wherein the fingerprints of extreme temperature can be found across many aspects of economic activity today.

For instance, researchers at University of Pennsylvania and Columbia University found that US automobile manufacturing becomes less productive on hotter days. Looking at weekly production data for dozens of plants from 1994 to 2005, they found that a week with six or more days above 90°F (32.2°C) reduced output that week by 8 percent or more. These are plants that collectively employ tens of thousands of blue-collar workers, making cars, SUVs, and light trucks in cities like Norfolk, Virginia; Georgetown, Kentucky; Detroit, Michigan; and Fremont, California.[21]

Here too, the researchers employed the natural experimental method to their advantage, looking at differences in weekly production within the same plant, and even controlling for seasonal variations by region of the country and type of automobile being produced (e.g., cars, vans, trucks), as well as the timing of new models going in and out of production.

While we cannot be certain that these effects are due to reduced productivity per se (as opposed to, for instance, increased worker absenteeism or correlated disruptions up and down the supply chain), the net effect is certainly suggestive of extreme temperature having a damaging impact on realized production. This seems consistent with reports that many manufacturing facilities and warehouses are too large—and the internal heat generated by production machinery too great—to cool cost-effectively on very hot days.

Moreover, the researchers found no evidence that companies were able to recover in the following weeks: plants were not more likely to schedule overtime in a week that followed bad weather, nor were they able to increase production.[22] The fact that they also found that high winds and rain reduced production suggests potential supply-chain disruptions; this is something that has been substantiated in recent work looking at international propagation of climate shocks across firm-to-firm networks, as we will discuss in chapter 12.

Similarly, economists Phillipe Kabore and Nicholas Rivers found even larger negative impacts of extreme temperature on manufacturing productivity in Canada. Drawing on a database of over 39,000 establishments from 2004 to 2012, they found that both hot and cold temperatures reduced manufacturing output considerably—not just for cars but for a wide range of manufacturers, in sectors such as aerospace, chemicals, textiles, and pharmaceuticals. Collectively, the firms in their study employed over 200,000 workers.[23]

A single day with daily mean temperature above 24°C (75°F), which one might think of as a day with maximum temperatures in the high 20s (°C) (80s [°F]), reduces annual output by 0.11 percent. One way to think about this is that it corresponds roughly to a same-day productivity decline of approximately 28 percent relative to a milder day in the 12°C to 18°C (54°F–64°F) range.[24]

It is worth noting that the threshold for heat-related effects appears to be lower, and the negative effects larger, than in the US study. The researchers also found that extreme cold reduced production as well, but only at very frigid temperatures. A day with mean temperatures below −18°C (0°F) reduced annual output by 0.18 percent, but less extreme cold days did not appear to have a meaningful impact. It seems Canadians are quite hardy when it comes to the cold, but less so for the heat. There is an active area of research aimed at understanding whether this is a general pattern, and if so, why, which may help us better adapt to a warmer world, and which we will explore in more depth in parts III and IV.

The United States and Canada are two of the richest, most highly industrialized countries in the world.[25] What then, might we expect in poorer, developing countries?

Emerging evidence is sobering. Looking at individual worker-level data from manufacturing firms in India, a team of economists at the

University of Chicago and elsewhere found that productivity fell on hot days by 2 to 4 percent *per degree Celsius* above room temperature. The implication is that a day with mean temperature above 25°C (76°F) reduced same-day output by approximately 32 percent relative to a milder day. They found these effects in cloth weaving, garment sewing, and steel infrastructure product manufacturing.[26]

Because they are looking at worker-level data, the researchers can rule out absenteeism as a potential driver, wherein output declines on hotter days because workers fail to show up to work. In fact, they found that while some salaried workers were less likely to show up during heat waves, piece-rate workers (who are paid only for what they produce and comprise the majority in their setting) were not. The latter category of workers—who constitute the majority in this setting—appear to toil through hotter working conditions, being paid less because they manage to produce less.

Another recent study looking at Chinese manufacturing also found negative effects of extreme temperature on output, to the tune of −0.56 percent annual production for each day during the previous year with daily mean temperature exceeding 90°F (32.2°C), implying that work effectively halts on such days, and possibly even creates negative spillovers on the productivity of other days that week, month, or year.[27]

The point I'd like to emphasize is that hotter temperatures may *already* be affecting companies' bottom lines—not 100 years from now, but yesterday, today, and tomorrow—and that, while many of these effects may not be catastrophic in any single instance, their cumulative effect could add up, especially in a warming world. Moreover, if the effects just described generalize even modestly to settings like construction or learning, which represent investments in future economic growth, the compound effects over time may be much larger than first meets the eye.

It also suggests that, were we better able to account for the hidden costs of adaptation—including those adaptations already in place—the total economic burden of extreme temperature may be even larger. If, for instance, the effect of heat on productivity in the US South is lower than that in Northern Canada because manufacturers in Georgia have invested in multimillion-dollar ventilation and cooling systems while their Manitoban counterparts have not, shouldn't we consider those additional costs in thinking about the economic consequences of a hotter world? The same could be said about stranded adaptation investment: in a warming world, whatever adaptations Canadians have managed to make them more resilient to the cold may be worth less and less over time.

Geography, Temperature, and Income Revisited

Bigger data married with increasingly refined tools of causal inference are two important ingredients that help illuminate previously obscured patterns. In particular, they show just how subtly pervasive the negative effects of hotter temperature can be in the real world, whether in the form of impeded learning, elevated workplace injury risk, higher rates of violent crime, lower firm-level output, or lower worker productivity.

Many of the effects documented are, in any given instance, imperceptibly small—a few percentage points here and there. But the bigness of the data and the intricacy of the causal inferential machinery allow us to identify even such small effects with increasing precision. Because these small cuts are spread across many millions—sometimes billions—of students, workers, and companies, the societal costs could add up quickly.

In particular, an important difference between recent work and previous work on climate damage is what one might call *granularity* in impacts. They are decidedly more micro than macro in the sense that there is an emphasis on trying to better understand the specific channels through which a given facet of climate change (e.g., extreme heat, flooding) may affect individual level outcomes such as health, productivity, or schooling.

Many early estimates of climate damages, including the influential dynamic integrated climate-economy (DICE) model, assumed for instance that manufacturing would be unaffected by climate change because it occurred indoors. New evidence suggests that this is far from the case, and that extreme temperature and precipitation events routinely reduce productivity in manufacturing plants, even in highly developed countries like Canada or the United States, though to lesser degrees than in poorer countries. In fact, one recent study suggests that extreme temperatures significantly impact earnings in over 40 percent of industries.[28]

These findings should influence how we think about climate change. Increasingly, they are being incorporated into policymaking. In 2020, the US government reconvened the Interagency Working Group on the Social Cost of Greenhouse Gases. A major goal was to update the social cost of carbon—a policy parameter that plays an important role in determining which government regulations can move forward—so that it might incorporate the emerging insights from this kind of research. As we will discuss in chapter 11, this update alone may have important implications for how the United States, and many other countries that typically follow US leadership on the social cost of carbon, approach climate policy.

If these findings help us think about our future on a warming planet, could they also help us understand our past? Let's use figure 6.4 to revisit the cross-country correlation from above.

If hotter temperature can influence health and productivity on so many levels, could it be possible for some part of this correlation to be causal, driven by the cumulative influence of a hotter climate on economic productivity, investment, and growth? Could at least part of Nigeria's relative poorness be because it is just that much hotter than Norway, and has always been?

While the data discussed in this chapter provide new clues, they aren't enough to settle this debate conclusively. There are still too many unanswered questions, too many real-world complications that cannot be brushed under the rug; one of these is the issue of adaptation, particularly over longer periods of time.

And yet, the rising tide of emerging findings is certainly intriguing. The new data suggests more strongly than ever before that, while institutional and other differences between rich and poor countries may be first-order determinants of economic growth, hotter temperatures may throw enough sand in the gears of human performance and productivity to drive at least some portion of the relationship described above.[29]

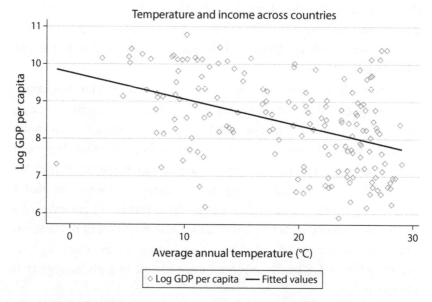

FIGURE 6.4. The relationship between population-weighted average annual temperature and log GDP per capita by country in 2000. *Source*: Author's calculations using data from Melissa Dell, Benjamin F. Jones, and Benjamin A. Olken, "Temperature Shocks and Economic Growth: Evidence from the Last Half Century," *American Economic Journal: Macroeconomics* 4, no. 3 (January 2012): 66–95.

In 2010, a study published in the *Proceedings of the National Academy of Sciences* made a bold, and at the time highly controversial claim: hotter temperature might routinely be reducing economic output. It made the argument that, contrary to the popular emphasis at the time on agricultural output, the effects of climate change on labor productivity and economic activity generally may be just as damaging, if not more so.

The study, conducted by Solomon Hsiang, found using data from twenty-eight Caribbean economies that a 1°C hotter-than-average year reduced non-agricultural output by 2.4 percent.[30]

It is worth putting this number in perspective. For a country like the Dominican Republic, whose GDP is roughly $110 billion, the suggestion is that a 2°C hotter-than-average year causes non-agricultural damages that year of roughly $5 billion. For the Caribbean region, which includes Cuba, the Dominican Republic, and Puerto Rico, the effect would be on the order of $20 billion in damages from one year that was 2°C hotter. These damages are on par with the $18.5 billion in (arguably more visibly salient) damages caused by Hurricane Rita, a category 5 hurricane that ravaged Cuba, Texas, and Louisiana in 2005, or the $19 billion caused by Hurricane Laura in 2020.

At roughly the same time, another highly influential study was published in the *American Economic Journal: Macroeconomics* by economists Melissa Dell, Ben Jones, and Ben Olken (whom I will refer to as DJO from here on). It found similarly negative effects of hotter temperature on economic output, not limited to agricultural impacts. Using GDP data from 138 different countries over several decades, they found that hotter temperatures appeared to reduce the rate of economic growth, at least for the poorer half of the world. A 1°C hotter-than-average year reduced GDP growth by 1.3 percent in these poorer countries.[31]

If indeed hotter temperature reduces GDP *growth*, then the consequences could be profound. For perspective, a country that grows at 4 percent a year will, over the course of five decades, achieve an income level 62 percent higher than that of a country that starts similarly but grows at 3 percent a year.

DJO provided further evidence that these were indeed effects on GDP growth, noting that the effect appeared to persist even when looking at longer time horizons. When they looked at changes in average growth rates over a decade or more with the change in average temperatures over the corresponding decade plus, they found a similar impact.

I remember reading these papers with fascination as a student. I also remember the vehement criticism they received. Many attacked the findings as a black box and claimed that the numbers were too big to be believable, especially given limited evidence regarding the potential mechanisms.

Fast forward a decade, and scientific evidence increasingly indicates that these studies may have been prescient. The links between nonoptimal temperature and productivity have now been shown across a wide range of settings, both in the lab and in the field, using macro- and microeconomic data. As we've seen, these include studies of manufacturing plants in the United States, Canada, India, and China, and students taking exams in New York, Israel, and Beijing. They also include not only tennis players, but also chess players and baseball umpires, all of whom appear to be nontrivially affected by temperature.[32]

The new data reveal a new twist: these effects, which at the time of Hsiang's and DJO's studies were limited to mostly developing countries, are pervasive across rich and poor countries, just much smaller in magnitude and more concentrated in vulnerable subpopulations within richer countries.[33]

As the adage goes, "absence of evidence is not evidence of absence." Both the Hsiang and DJO studies were largely macro in the data that they used. New, more granular data, much of it subnational (and, in many cases, using individual-level outcomes like worker productivity, student performance, and firm profits), suggests that the lack of detectable effect in rich countries may have been an artifact of two factors. The first factor is that a hotter-than-average year may mean a milder winter in higher latitudes, which could have beneficial effects for colder countries. Second, in many rich countries even hotter years that were the result of hotter, possibly more damaging summers may have resulted in small average effects given relatively high incomes and correspondingly high levels of adaptation whose costs had either been paid in prior years or were being mixed in with the overall changes in output.

We now have a strong evidentiary basis to believe that, for those individuals within rich countries who *do not* have access to such defensive investment, hotter temperatures can have large negative effects. We will explore the implications for our understanding of climate inequality and environmental justice in part III.

So how large are the potential macroeconomic consequences overall? One recent estimate suggests that 2°C of warming in the United States could reduce economic output by approximately 1 to 3 percent of GDP each year, which even for the economy today would amount to roughly $420 to $630 billion.[34] To put this in perspective, the ten most profitable American companies had a combined annual profit in 2021 of $520 billion.[35] So, with 2°C of warming, the effects of hotter temperature on the United States would, each year, be akin to wiping out the profits of Apple, Berkshire Hathaway, Alphabet (Google's parent company), Microsoft, JP Morgan Chase, Meta Platforms (Facebook's parent company), Amazon, Bank of America, Exxon Mobil, and Fannie Mae . . . combined.

One important disadvantage of methods like these is that using high frequency weather variation to identify causal effects of climate shocks is that extrapolating over the longer term involves important uncertainties. For instance, the effect of a 2°C hotter year may be an overestimate of the effect of a world that is permanently warmer by 2°C if there are forms of adaptation that may become available in the longer term but are not available in the short term. These may include innovations in crops that make them more heat-resistant, cheaper, and more effective cooling devices, or better early warning systems that reduce the number of hospitalizations. Conversely, it is possible for adaptation strategies that are available in the short term to become unavailable over time, leading such weather-based studies to *underestimate* the longer-term effects of climate change. One such possibility is that, as communities such as those in the American West begin to draw down local aquifers, adaptation strategies that relied historically on the availability of cheap water may no longer be accessible. A vibrant research community is actively seeking answers to these puzzles.

Because there are so many potential means of adapting to heat's harmful effects, whether in the form of air-conditioning, changes in public institutions and policies, or changes in individual behavior and lifestyles, understanding what climate change means for the world economy is fraught with challenges.

One thing seems clear, however. Hotter temperature affects economic productivity in subtle but potentially profound ways, and climate change, without remedial investments in adaptation, has the potential to significantly alter the economic playing field.

7

Peace and Peace of Mind in a Hotter World

Daily existence is very different if you always have to worry about being abducted, raped, or killed.

—STEPHEN PINKER, *THE BETTER ANGELS OF OUR NATURE*

IN A REFUGEE CAMP on America's doorstep, hundreds of residents line up at makeshift laundry stations to handwash their muddy clothing. As one observer describes it, "A thick heat baked the rotten debris, a mixture of broken toys, human waste and uneaten food swarming with flies . . . The heat just kind of hangs over you."[1]

One of several tent camps along the Mexican side of the US–Mexico border, these temporary tenements serve as a sort of Dante's limbo for migrants seeking entry into the United States. Many are asylum seekers—families with women and children, young and old—hailing from Honduras, Guatemala, and other parts of Central and Southern America. Some are fleeing persecution and violent conflict. All are seeking refuge in the United States.

Amnesty International defines an asylum seeker as a person who has left their country and is seeking protection from persecution or human rights violations in another country, but who hasn't yet been granted asylum.[2] In this case, many have reportedly been victims of gang violence, including extortion attempts and kidnappings. Some have had loved ones killed.

Ruth Monroy and her family faced increasing extortion payments on their bakery in El Salvador. So they fled north, along with their three children, ages 8, 6, and 3.[3] "We're just humble people looking for protection, for peace," she explains. For many families like Ruth's, the stakes could not be higher. "If the US sends us back, they might as well send five coffins with us. We'll be killed as soon as we step off the plane," says Ruth's husband Juan Carlos.

The possibility that climate change may increase the flow of cross-border migration has raised increasing concern. "A great upheaval is coming," warns one report. "To survive climate breakdown will require a planned and deliberate migration of a kind humanity has never before undertaken." As Steve Killea, president of the Institute for Economics and Peace, puts it, climate change "will have huge social and political impacts, not just in the developing world, but also in the developed, as mass displacement will lead to larger refugee flows to the most developed countries."[4]

Some migration episodes are the product of full-blown humanitarian crises. Examples include the Syrian refugee crisis of 2015, and the Darfur conflict, which has been labeled by some observers as "the first climate change war." In 2007, then–UN secretary general Ban Ki Moon said of the conflict that "among diverse social and political causes, it began as an ecological crisis, arising at least in part from climate change."[5]

Will climate change increase the flow of asylum seekers like Ruth and her family? Critical in assessing how such a chain of events might unfold is the link between climate change and circumstances in the sending countries. This includes the potential link between climate change and violent conflict.

A growing body of evidence suggests that climate change may indeed already be increasing the propensity for violent conflict and thus possibly already increasing the rate of international migration.[6] And yet, a closer look at the evidence suggests that for every prominent humanitarian crisis like the so-called migrant caravan from Central America to the United States, or the war in Darfur, climate change could also already be creating subtler and perhaps more widespread changes in the way human beings relate to one another, and perhaps even to ourselves. The tools and evidence explored so far suggest there may be more to the story than first catches the eye.

Temperature and Violent Crime

Safety, like water or air, is a latent attribute that does not present itself as vital until it is lacking. This is obviously true for physical safety, which we might define as freedom from bodily harm. It is arguably also true for the psychological aspect of safety, whether in the form of verbal abuse or the threat of physical harm.[7]

One indication that we value public safety is that societies appear willing to allocate substantial sums to obtain and maintain it. In 2019, state and local governments in the United States spent $125 billion on policing, $82 billion on corrections, and $50 billion on the courts, amounting to approximately 7 percent of their total direct expenditures. For context, this is somewhere between total expenditures on highways and roads (6 percent) and higher education (9 percent).[8]

America's criminal justice system may be especially resource intensive, but it is not the only country that spends a considerable manpower and tax revenue to maintain safety and public order. The European Union (EU) as a whole spent roughly €250 billion, or €500 per person on public order and safety in 2015. Countries like Luxembourg (€900), the United Kingdom (€790), and the Netherlands (€720) spend substantially more than that per person.[9]

Of course, to the extent that individuals also spend their own time and resources to ensure personal safety, whether in the form of home surveillance devices, martial arts lessons, or pepper spray, the aforementioned public expenditures represent an incomplete account of societies' willingness to pay.

Perhaps a more direct way to measure safety—or lack thereof—is to look at the incidence of violent crime.

Rates of crime vary enormously both across and within countries. As violent as a gun-saturated America can seem, rates of homicide are far higher in countries like Namibia, Brazil, Mexico, and South Africa. In the United States, roughly 6 people per 100,000 are murdered each year. That rate is closer to 12 per 100,000 in Namibia, 22 in Brazil, 29 in Mexico, and 36 in South Africa. At the top of the most homicidal countries list is El Salvador, where over 50 out of every 100,000 individuals are murdered each year.[10] To put that in perspective, in the United States, roughly 50 out of 100,000 people die from Alzheimer's and diabetes combined, suggesting that El Salvadorians are just as likely to be murdered as Americans are to die of diabetes or Alzheimer's.[11]

In contrast, homicide rates in countries like the United Kingdom, Germany, South Korea, and Japan are a fraction of this, at around one or less per 100,000.[12] And a much smaller share of these involve gun violence.

Even relatively less affluent countries can achieve higher levels of safety, at least with respect to homicidal violence and according to official crime statistics. Malaysia has a homicide rate that is a fraction of that in the United States, despite having much lower incomes per capita. Of course, we must note that these statistics, as with any official statistics on crime, rely on records of crimes that were reported, which may diverge from what actually occurs depending on many factors, including whether individuals feel comfortable making reports and whether local law enforcement has the capacity to investigate. Here, the data is taken from the United Nations Office on Drugs and Crime. It seems difficult to attribute such enormous differences to measurement issues alone.

For all the reasons noted in the previous chapter, it seems misguided and insensitive to suggest that living in a hotter climate makes the culture more prone to violence, especially given the colonialist undertones of early musings on this subject, which often intimated a link between tropical climates and lawless behavior.[13] And here too, looking at the cross section is dangerous. It does not take complicated analysis to make the casual (not *causal*) observation that hotter, poorer, places tend on average to have higher rates of crime. But a hotter climate is by no means the only factor that could be responsible for such differences.

Hopefully by this point in the book statistical alarm bells are going off in your mind. This is the omitted variable bias in another guise. To the extent that hotter climates are also correlated with greater incidences of poverty, weaker institutional capacity, or other potential determinants of violence, one would be wise to exercise caution in making causal claims.

The trends in data availability, computing power, and causal inferential tools noted previously have helped shed valuable light on this question. In one study, Matthew Ranson used data from over 17,000 US law enforcement agencies across 2,997 counties to estimate the effect of extreme temperature on various criminal offenses, including murder, manslaughter, rape, and aggravated assault.[14]

Looking at daily local temperature data from 1980 to 2009, he found that, across a variety of offenses, higher daily temperatures lead to more crime that month. A week of temperatures above 90°F (32.2°C) raised the monthly rate of rapes by over 5 percent, and increased murders and aggravated assaults by around 3 percent. Once again, we can be more confident regarding the causality of this association because the researcher is looking at variation in both temperature and crime rates within a given county over time, as opposed to comparing crime rates in hotter versus cooler places.[15]

These are nontrivial effects. Ranson estimated that the increase in hotter days brought about by climate change could lead to over 1 million additional cases of aggravated assault and nearly 200,000 additional cases of rape in the United States over the course of the twenty-first century. Noting that other research finds that a 1 percent increase in the size of a city's police force results in a 0.3 percent decrease in violent crimes, he argued that an immediate and permanent increase of at least 4 percent in the size of the US police force would be required to offset the crime increasing effects of climate change.

These findings are echoed in many more recent studies, including one by economists Kilian Heilmann and Matthew Kahn. Using data from the Los Angeles Police Department (LAPD), they found that on days with maximum temperatures above 85°F (29.4°C), violent crime increased by 9 percent compared to cooler days.

By using administrative data from the universe of LAPD call of service records, the researchers were able to explore the temperature–crime relationship with even more granularity, down to the district (of which there are many hundreds within the county of Los Angeles) and the day of the crime. The added resolution is helpful in showing that, for the most part, it is temperature on a given day that leads to changes in crime that day, as opposed to a lagged effect.

As we will see, this provides important additional evidence regarding the potential mechanisms behind the relationship. The more granular data is also helpful in highlighting just how much the effects are concentrated in lower-income neighborhoods, even within the same city of Los Angeles. These highly localized vulnerabilities may be critical in understanding the implications of climate change for inequality and economic opportunity at the individual level, and possibly also what can be done about it.

A growing body of evidence indicates that hotter temperatures increase the likelihood of violent crime. On one level, this fact is concerning on its own, as it portends more violent conflict as the world warms. At the same time, it still begs the important question as to *why*—the answer to which may be critical in informing our understanding of the possible extent of climate change's potential societal consequences.

Why does hotter temperature lead to more violent conflict? At least three potential pathways have been proposed. One hypothesis is that hotter temperature leads to hotter tempers, also known as the heat-aggression hypothesis. The emphasis here is on the link between temperature and our emotional circuitry, whether in the form of increased negative affect, reduced inhibition, or something else. Another posits that changes in weather conditions may influence the costs and benefits of engaging in criminal activity. This could be because heat and drought reduce livelihoods, and deteriorating economic conditions give way to more acts of desperation, sometimes of the violent variety. Or, it could be that inclement weather leads to laxer law enforcement, thus reducing the perceived costs of a given criminal act. Let's call this the *changing incentives mechanism*.

Finally, it has been observed that temperature may affect the frequency of social interactions. This could be because colder, wetter weather tends to keep people indoors, and sunnier, warmer weather brings more people out into the proverbial town square. This hypothesis stresses the potential mechanical connection between weather conditions and the number of interpersonal interactions, some proportion of which may be conflictual, perhaps especially so when people who don't usually meet must mingle. We'll call this the *added interactions mechanism.*

Which of these mechanisms are at play matters, in part because each pathway has different implications for how we prepare for a warmer world. This is especially true if one is interested in designing policies that weaken the link between warming, violence, and the many social costs that such violence engenders.

It also matters for our understanding of the magnitude of climate impacts more broadly. Of particular concern is the possibility that hotter temperatures influence human temperament, thus affecting the nature of our social interactions across the board—far beyond the relatively few incidents that result in outright physical violence or a rise in international asylum seekers.

Road Rage

In a fascinating 1986 study, researchers Douglas Kenrick and Steven MacFarlane set out to test a simple hypothesis. Does hotter temperature increase expressions of road rage?[16]

Over a four-month period beginning in April, researchers in Phoenix strategically located themselves at a one-lane intersection in a residential area of the city. Every Saturday between 11:00 A.M. and 3:00 P.M., they sent a collaborator, whom we will call Dana, into the fray, putting her behind the wheel of a 1980 Datsun 200SX. Dana would park her vehicle near the target intersection, and when the light turned red, she would drive to the head of the intersection and wait for an unwitting subject to pull up behind her.

When the light turned green, Dana intentionally remained stationary throughout the course of the green light, which lasted for twelve seconds. In order to maintain as neutral a position as possible, she was instructed to "keep still, with her eyes forward, car in neutral, foot off the brake, and her hands on the steering wheel." Only when the green light had ended did the collaborator make a legal right turn to exit the intersection.

The researchers observed each of these trials from a close distance, though hidden from view. Presumably behind some large cacti, perhaps like Wiley Coyote spying on an unwitting Roadrunner. At the onset of each trial, they counted the number of horn honks delivered by the subject (the driver who had pulled up immediately behind) during the twelve-second green light. They also estimated the duration of each honk, and measured the time it took until the first honk. In addition, they recorded whether the car windows were open or closed, the gender of the driver, the number and presumed genders of other passengers, and the number of cars behind the subject. They then matched these observations with local hourly temperature and humidity readings. The temperatures during the time of the interactions ranged from 84°F (28.9°C) to 108°F (42.2°C).

The researchers found that, for the twenty-nine subjects who had their windows down, outside temperature was strongly correlated with both the number of horn honks and the time it took for the first horn honk to occur. Hotter temperatures meant more frequent horn honks that took fewer seconds to materialize and that lasted longer once they did. At a humidity-inclusive heat index of 105°F (40.6°C), drivers were recorded as spending over twice as much time honking relative to when heat indices were between 86°F and 90°F (30°C and 32.2°C). (Among the list of variables that were not measured in detail but one wishes could have been recorded might be the amount of profanity used by the impatient drivers.)[17]

The researchers surmised that these hot-headed subjects did not have air-conditioning in their vehicles and were feeling extra impatient when they were already hot and bothered. For the forty-four subjects who did not have their windows down—and presumably had air-conditioning running—there was no statistically significant correlation between temperature and horn honking.

In case you were wondering who in their right mind would drive a car without air-conditioning in Phoenix, Arizona, I wondered the same thing. It turns out that as of 1980 roughly 70 percent of all new cars in the United States had air-conditioning. If we assume, perhaps conservatively, that the average lifetime of an automobile was somewhere in the order of five to ten years in those years, it stands to reason that a nontrivial fraction of cars on the road during the study did not have air-conditioning installed. Alternatively, it may have been that, at an inflation-adjusted price of roughly $4 per gallon, significantly lower gas-mileage and a per capita income that was roughly a quarter of what it is today, a fraction of drivers might have been trying to save fuel by leaving the air-conditioning off.[18]

Studies like these are instructive but incomplete. While they are an improvement on laboratory studies in that they come from real-world interactions, their external validity may be hampered by the fact that the kind of hostility documented in this setting may not be as relevant in other settings. Driving is both a highly common but highly peculiar activity. As the researchers note, "The automobile provides its owner with protection, unusual power, easy escape, and some degree of anonymity." We have no doubt witnessed at some point in our lives a disconnect between the level of offense and the severity of road rage metered out in response. This is one reason for potentially limited external validity.

Another has to do with gender. Recall that the driver being honked at was always female and she was instructed not to engage. At the risk of speculatively backward-casting unfortunate gender stereotypes, one cannot help but wonder whether the same relationship would hold if the researchers had recruited a physically imposing body builder to be the experimental driver, or if the car at the light was a police cruiser.

Perhaps just as importantly, it is unclear to what extent the drivers who had their windows down on hotter days differ from other drivers on the road. This is an issue of selection bias. It seems plausible for there to be a connection between socioeconomic status and the likelihood of driving without AC on a hotter day, and for this connection to be partially responsible for the relationship between temperature and horn honking behavior noted above, if for instance we have reason to believe these individuals may be under more baseline time or financial stress. Without comparing the same driver to herself on hotter versus cooler days, it is difficult to discern.

And yet, this study provides some interesting food for thought. Might hotter temperatures make us more irritable? Do our tempers flare when the mercury rises?

Heat on the Mound and in the Red Zone

Evidence from athletics matches provides further clues. In a study of major league baseball, Alan Reifman, Richard Larrick, and Steven Fein found hotter temperature to be associated with a significant increase in the number of batters hit by a pitch during the game. The relationship appeared to hold even when controlling for the total number of wild pitches, walks, and fielding errors, all of which may indicate a lack of ball control rather than increased aggression.[19]

In a study of the National Football League (NFL), researchers found that hotter game-day temperature was associated with an increase in the number of aggressive penalties.[20] In case you were wondering whether football games, which are typically played in the autumn and winter months (unlike baseball, which is mostly a summer sport), are ever played in hot enough conditions, the hottest game-time temperature in their data was 109°F (42.7°C). Looking at game-level data from all NFL seasons between 2000 and 2011, they found that penalties like taunting, face masks, unnecessary roughness, and unsportsmanlike conduct increased during games played in hotter conditions.

Interestingly, they did not find a significant increase in the number of false starts, pass interferences, or other penalties not obviously associated with increased aggression. Furthermore, the relationship between heat and aggressive penalties was only significant for the home team. Opposing teams did not appear to commit significantly more aggressive penalties on hotter days.

Both sets of findings suggest that hotter temperatures may indeed have some connection with human aggression. But by focusing on players' behavior on the field, we have replaced one set of external validity concerns with another. It is difficult to infer from these findings whether such a relationship exists in less obviously competitive environments, like a conversation over dinner or an important sales call.

Temperature, Temperament, and Twitter

Social media offers unprecedented access into the inner musings of billions of people around the globe. Whether we are aware of it or not, by posting on the web from cellular devices, we are creating, among other things, a temporally and spatially referenced map of our expressed thoughts and opinions, likes and dislikes, perhaps even our internal emotional states.

Economist Patrick Baylis sought to use data from X (formerly Twitter) to forensically discern whether temperature meaningfully affects our emotional sentiment, also referred to as *emotional affect*.[21]

To explore whether temperature meaningfully affects our daily sense of satisfaction, Baylis compiled over a billion tweets from across the United States and six other countries. That's billion, with a B. Using natural language processing algorithms designed to extract emotional state from unstructured text data, he was able to categorize tweets according to a quantitative measure of expressed sentiment.

Many previous studies of subjective well-being used survey data. A common challenge in these types of studies is that measurements of subjective well-being are not only subjective across individuals but also may include hidden differences across time and space. The way one responds to questions about well-being may depend on regional cultural norms or linguistic dialects and may also be colored by the interaction between the interviewer and the interviewee, which may itself be a function of temperature. By gathering circumstantial data on tweets, researchers like Baylis are able to circumvent many of these problems. Such data have the benefit of representing expressed sentiment without the potential for observer or Hawthorne effects, and they have the added benefit of providing massive data at relatively low cost, at least compared to surveys.

PEACE AND PEACE OF MIND IN A HOTTER WORLD   137

However, even with such great data, to answer this question convincingly, important (and by now hopefully familiar) causal inferential hurdles needed to be overcome. Using regional variation in sentiment may be problematic if individuals with higher incomes tend to experience higher levels of life satisfaction and can also afford to locate in areas with generally pleasant climates. Similar to comparing crime rates in cooler Minneapolis to warmer Miami, comparing the frequency of profane phrases in tweets across cities could lead to a biased relationship between weather and expressed sentiment. Using seasonal variation in sentiment might mistakenly attribute to temperature things that are changing with the seasons, like seasonal unemployment, more students being out of school and bored during the summer, or unusual traffic jams during holiday weekends.

As with many of the other studies we have discussed so far, Baylis exploited natural variation across days within regions. Specifically, he looked at daily temperature variation within a metropolitan area to account for any regional variation as well as within month and year, to account for differences in sentiment across seasons and years.

He presented some compelling findings. On days with maximum temperature above 102°F (39°C), measured sentiment decreased by approximately 15 percent of a standard deviation.[22] How big is this effect? Compare this to the average difference in expressed sentiment over the course of the week where the difference between Sunday and Monday appears to be roughly 10 percent of a standard deviation on average. In other words, a 100°F (37.8°C) day dampens one's mood more than moving from the weekend to the start of the workweek.

Interestingly, Baylis found that hotter temperature led to a significant increase in the use of online profanity. A day with temperatures above 102°F led to a 3–4 percent increase in profanity. In particular, such days lead to a 5 percent increase in "aggressive profanity," which he defined as tweets that included a popular, aggressively profane phrase.

But wait; what if there was a similar selection problem at play here as well? Perhaps there is a correlation between the users who choose to tweet on hotter days, not unlike the potential correlation between drivers who chose to drive with their windows down on hotter days in Arizona?

The bigness of the data allows us to account for this possibility as well. Looking at variation in tweet sentiment and temperature across days within a given user, Baylis was able to control for potential biased selection into and out of the sample. The results from this analysis were slightly less precise because he was limited to more active users, but nevertheless confirmed the same qualitative pattern. The fact that Baylis documented these patterns not only in a single city in the United States, but across the entire country suggests that this is not a phenomenon limited to especially irritable or moody tweeters.[23]

These and other studies suggest that elevated temperatures affect our temperament for the worse across a range of settings. One implication is that, as the world warms, our relationships with others may be just that much more likely to tip into conflict, verbal or physical.

The evidence also increasingly indicates that hotter temperatures may adversely impact our relationship with *ourselves*.

Approximately 18 percent of Americans reported struggling with a diagnosable mental, behavioral, or emotional disorder in 2018.[24] And the societal costs associated with mental health issues appear to be growing. In the United States, the direct costs of treating mental health disorders surpassed $200 billion in 2013 and were growing at 5 percent annually even before the COVID-19 pandemic, which likely increased them significantly. The indirect costs, which include lost productivity, are likely much larger.[25]

A fascinating study by economists Jamie Mullins and Corey White finds that hotter temperatures may exacerbate such mental health disorders significantly. Looking at data from many millions of inpatient records in the United States, including over 5.9 million emergency department visit records for the state of California (2005–2016), county-by-month suicide statistics for over 1.6 million suicides across the United States, and self-reported mental health status surveys for over hundreds of thousands of individuals across the United States (1993–2015), they found that increases in temperature lead to increased emergency department visits for mental illnesses. A 1°F increase in monthly mean temperature increased the monthly emergency department visit rate by 0.4 percent.[26]

Hotter temperature also appears to increase self-reported days of poor mental health and even leads to a significant uptick in suicides. Mullins and White found that, in the United States, a 1°F increase in mean monthly temperatures increased monthly suicide rates by 0.3 percent.[27]

These findings are consistent with other evidence, including work by economist Tamma Carleton, who found that hotter, dryer weather led to a significant increase in suicides in South Asia; although there, it appears likely that correlated reductions in agricultural livelihoods may be an important additional driver of the heat–mental health relationship.[28]

The pattern that emerges from these studies is a disconcerting one. In addition to many tangible effects on human life and economic productivity discussed in the previous chapter, hotter temperature may have a wide range of more intangible adverse consequences for our mental health and overall quality of life. If measurement issues associated with traditional data sets were a problem in the context of heat and mortality or heat and labor productivity, they are likely even more problematic here, as many of these impacts are almost impossible to document on a case-by-case basis. There are echoes here of the chapter on human capital, where we noted the significance of nonmarket climate costs that typically aren't captured in official statistics.

Only when looking across larger populations—and here too, carefully controlling for a wide range of potential spurious correlations—can we detect with confidence the hidden footprint of heat on our minds.

Back to the Border

One of the hypothesized potential links between climate change and cross-border movements involves the possibility of increased violent conflict. If climate change increases the likelihood of conflict in one part of the world, it stands to reason that one consequence may be increased pressure on other countries' borders.

A growing literature suggests that, both across and within countries, hotter temperatures are causally associated with increased violent conflict, including not only interpersonal crime but intergroup conflicts as well.[29] While it is challenging to attribute any given conflict to climate change, quasi-experimental evidence indicates that both hotter and drier conditions may increase the likelihood of civil and international conflict. There is evidence now suggesting that extreme weather events in Mexico increase migrants to the United States, and that asylum applications to the European Union respond significantly to weather events in sending countries.[30]

Perhaps this is one of the reasons why the military is increasingly interested in climate change as a potential source of geopolitical instability. The US Department of Defense (DoD) has elevated climate change as a national security priority, noting its potential implications for "geostrategic, operational, and tactical environments" in which US national security and defense forces must operate. As it states in a 2021 report: "To train, fight, and win in this increasingly complex environment, DoD will consider the effects of climate change at every level of the DoD enterprise."[31]

Will climate change increase unrest and violent conflict? If so, an important question for policymakers will be the mechanisms that explain why, and whether the initial conditions can be acted upon before violence erupts.

How much of a climate-conflict relationship will be driven by deteriorating economic conditions, such as drought and famine? To what extent could the availability of weapons, or of idle hands due to high youth unemployment exacerbate conflict due to climate change? How much of future cross-border migration will be driven by tumult and unrest in countries hardest hit by climate shocks?

These are important questions, especially to the extent that one is interested in reducing such drivers at their source. The answers to these and other questions regarding the mediating factors in the relationship between climate and conflict remain unclear. What *is* clear is that the news reels will tend to fixate on drama at the border, and in doing so will likely miss the bigger picture.

It's not only that the arrival of desperate migrants at the border— whether in the United States or the EU or elsewhere—is but a visible symptom to a hidden and as yet unconfirmed cause, often in the sending country. It's also that such coverage, by focusing on the migrants who have made it to the border, obscures the fact that the numerical majority of climate-induced refugees are likely to be those who do not have the wherewithal to make it to an international border.

The image of desperate climate refugees crossing international borders in the wake of natural disasters commands headlines. But this masks at least two important realities. One is that the vast majority of human migrations—including those set in motion by climate change—appear to be *intra*national, as opposed to *inter*-national. The second is that many relocations are temporary. Available evidence indicates that many displaced populations return to the disaster zone shortly after and rebuild right where they were before the disaster hit. One important caveat is that it is difficult to know with much clarity what will happen in the future when natural hazards such as hurricanes and floods become more chronic. Relatively few studies have been able to quantify the impact of repeated hurricanes and floods on the likelihood of out-migration.

Here too, the social and economic context may matter in important ways. In general, whether climatic pressures like drought or rising temperatures lead to out-migration depends in part on the economic capacity of the potentially displaced. It appears that the high fixed costs associated with moving, especially internationally, can be a significant constraint even in cases where pressure to move out is mounting. These costs may be pecuniary, in the form of shipping and transportation costs. They may also be cultural and social, in terms of relationships broken and cultural heritage lost.

In fact, recent studies find that gradually rising temperatures may actually *reduce* the rate of migration in many poorer countries. Using data from over 116 countries, economists Cristina Cattaneo and Giovanni Peri found that higher temperatures lead to increased rural-urban migration in middle-income countries, but to reduced rural-urban migration in low-income countries, consistent with this idea that lower-income households may face credit and liquidity constraints when determining whether to migrate.

A group of researchers working on climate-induced migration put it this way:

> There is a trade-off between the incentives to move and the resources to do so. This trade-off is particularly prevalent in the context of climate-related migration. On the one hand, poor people have higher incentives to migrate, as they tend to be those most exposed and vulnerable to the impacts of climate change, with limited capacity to adapt in situ. On the other hand, poor people often cannot afford to pay the cost of migration. Poorer people thus face a "double set of risks" by being both unable to move away from climatic threats and especially vulnerable to their impacts. Which forces prevail is an empirical question.[32]

It seems possible that climate change may, perhaps paradoxically, lead in some cases to greater *immobility,* as its impacts on livelihoods reduce the resources required to move.[33]

Evidence also indicates that a significant share of climate-induced migration arises from gradual onset events, in addition to the more salient crises. A recent metareview of over 116 studies finds that slow-onset climatic changes, in particular high temperatures and drying conditions, are more likely to increase migration than sudden-onset events, such as fires or floods.[34] It remains possible that the kinds of migrations that occur in response to sudden-onset disasters are typically more difficult to pick up in the data, suggesting that more research is needed to understand this feature of the debate more fully.

If climate change leads to a deterioration of economic conditions for the poor, and especially if the underlying economic growth rate of the poor is not sufficient to offset these added burdens, it seems plausible that climate change may be a self-compounding risk factor that keeps poor people trapped in gradually worsening environments.

As noted at the outset of this chapter, the potential for climate change to affect international migration is a highly salient feature of public debate. A 2019 headline in *The Economist* read, "How Climate Change Can Fuel Wars," with a drop head that read, "Droughts are already making conflict more likely. As the world gets hotter, mayhem could spread."[35]

PEACE AND PEACE OF MIND IN A HOTTER WORLD 143

Climate change may increase the flow of international migrants, including asylum seekers fleeing from violence, due in part to its effect on crime and conflict in sending countries.

Understanding the causes of these international migrations, whether they involve increased conflict or something else, are important priorities for global security and international relations.

But even here, we may want to check our impulse to fixate on international borders, which so often provide the focal point. For every family like Ruth's that makes it to the border, there may be countless others who are, if anything, increasingly immobilized by the adverse economic consequences of a changing climate, and many others still who will make their way to urban centers within their countries of origin, far from the sheen of the media spotlight.

A Temperamental Criminal Justice System

Let us return to the relationship between temperature and local violent crime. When a suspect of a crime is apprehended, they are typically tried in court. I say typically because there are still many parts of the developing world where the presence and proper functioning of such institutions as the courts and rule of law are not a given. Whether they are perfectly fair in their administration of justice in developed countries is also an open question.

In the United States, judges swear an oath to uphold the law, to adjudicate each case that comes their way based on the facts and nothing but the facts. Indeed, the Sixth Amendment to the US Constitution describes a "fair trial" as a fundamental right. The Administrative Procedures Act of 1946 states that any ruling or decision by an agent of the US government should not be "arbitrary" or "capricious."[36]

One can debate the merits of granting harsher or more lenient sentences for any given conviction. But what is perhaps less debatable is the idea that, for a given law, the adjudication of each case should be dispassionate and consistent with the law, based on the facts and the facts alone.

In some sense, the criminal justice system relies on the workmanlike precision and impartiality of its judges. There is the law. There are the facts. There is a ruling that reflects the impartial and consistent application of the law. Something as seemingly trivial as temperature on the day of a court adjudication could not possibly affect conviction and sentencing decisions, could it?

One analysis by economists A. Patrick Behrer and Valentin Bolotnyy suggested that, even in a rich, democratic country like the United States, temperature routinely influences judges' rulings. They found, using detailed court records from Texas, that hotter temperature on the day of sentencing leads to harsher sentences and a lower likelihood of cases being dismissed.[37] This is even though, given the hotter average climate of Texas, most homes and offices in the state have air-conditioning. As with many of the studies we've explored in this book, the researchers looked at differences in behavior for the same judge in the same location at different points in time, carefully controlling for other factors that might be correlated with temperature, but which might also affect judges' decisions.

These findings have been echoed in a number of settings, including among US parole judges, as well as among judges of Indian civil and criminal court cases, where researchers found in an analysis of 77 million Indian court cases that judges were more likely to issue a conviction on hotter days.[38]

What should we make of the relationship between temperature and judges' decisions? On the one hand, these results seem consistent with the idea that judges are human and can be influenced by their physiology and perhaps also their emotions. A recent study by Daniel Chen found that judges are actually more likely to rule favorably when a nearby home NFL team wins during the preceding weekend, perhaps buoyed by a surge of collective local euphoria. Unfortunately for many, the reverse appears to be true as well. One study in Louisiana found that juvenile court judges hand down significantly longer sentences following an unexpected loss by Louisiana State University's NCAA football team, and that those judges who are based closest to the university are most susceptible to this effect.[39]

On the other hand, they call into question the idea that institutions are monolithic constructs immune to the effects of climate. Indeed, seeing how the functioning of institutions, whether judicial systems, policing systems, or school systems, are ultimately dependent on decisions and actions of the many human beings who administer them, maybe we shouldn't be surprised that the subtle footprint of climatic factors can be felt in how they operate as well. Still, the possibility that climate change could, through its indirect effects on judges and others, subliminally tilt the scales of justice is one that I find concerning.

In the case of violent crimes such as assault, rape, or murder, a conviction most likely leads to incarceration. In the United States, over 2 million individuals are behind bars at any given point in time. Many of them have been accused of violent crimes. Given how rates of violent crime vary across cities and neighborhoods in the United States, it should come as no surprise that most of the incarcerated hail from high-poverty, high-minority populations.

Some sources suggest that many prisons do not have universal air-conditioning, including in very hot states like Alabama, Texas, Louisiana, Arizona, and Florida. This is even though, at least according to some reports, temperatures inside prison cells can reach well into the 100s (°F), with some having recorded temperatures above 140°F (60°C).

Should prisons provide air-conditioning for inmates and security personnel alike? Climate change injects newfound legal and policy relevance to questions such as these, particularly in light of recent findings. A 2021 study by economists Anita Mukherjee and Nick Sanders found that, on hotter days, violence in prisons spikes—both against prison security personnel as well as among the incarcerated. In Mississippi, a day with average temperature above 80°F (26.6°C) increases the rate of violent incidents by approximately 20 percent, compared to days with temperatures below that threshold. These include such incidents as physically assaulting and even murdering fellow inmates, as well as engaging in violence against security personnel.[40]

The reverberations of these conflagrations are often long lasting. Many of these inmates will, in the wake of such incidents, receive longer, harsher sentences.

It is worth reminding ourselves that, in work on heat and crime in the general population, the effect was found to be much larger in high-minority, low-income communities—even within the same city of Los Angeles. So as the world warms, it seems plausible that hotter temperatures may on the margin act as a catalyst for more violent crimes outside of prison, harsher sentences in the courtroom, and exacerbated sentences once in the prison system, and that some demographic groups will be more likely to experience such cycles than others.

Whether or not this changes the moral implications of the violent acts that are carried out is beyond the scope of this book. But one thing these studies suggest is that the way climate change ultimately manifests as human suffering may depend not only on its effects on physical systems like air masses and monsoons, but also its effects on social systems, including the courts, border control, and correctional facilities.

Another important aspect of climate change that they ask us to consider is the inequality: how the realized effects of a given climate hazard may depend on highly local, highly individual specific circumstances, whether that is the level of crime in one's neighborhood, the likelihood that the school or workplace or courthouse has air-conditioning, or the economic pressure points that are triggered by climate shocks such as heat or drought.

This in turn requires a nuanced understanding of the forces driving economic opportunity—both across rich and poor countries as well as between rich and poor people, wherever they may reside.

An adversarial framing of climate change science may make it seem like we need to attribute single salient events to humanmade climate change. This may be why a growing body of climate science is dedicated to *attribution science*, which uses increasingly sophisticated techniques to determine the likelihood that a specific heat wave, hurricane, flood, drought, or wildfire occurred or was made more severe due to human activities. The implicit theory of change appears to be that, by more strongly establishing the link between human activities and specific climate events, it reinforces the urgency of taking action to reduce greenhouse gas emissions to prevent a future in which such events become more frequent, making it difficult for skeptics to let humanity off the hook with talk of *natural variability*.

A deeper understanding of climate change requires us to ask what we might not yet know about the human consequences of a hotter climate, full stop—to perhaps pay closer attention to the hidden handicaps associated with a hotter climate irrespective of whether the added severity can be attributed to anthropogenic emissions or not. This is especially true to the extent that one is interested in understanding how such handicaps may affect the lives and livelihoods of the world's poor.

The evidence increasingly suggests that, even before we started changing the earth's climate, the subtle disadvantages of living in a hotter climate may already have posed a significant handicap to economic growth and human flourishing in many tropical countries. This makes an understanding of what kinds of adaptations are most effective at unshackling poorer individuals and nations from these and other climate fetters even more important moving forward.

PART III

A MAN IN A NAVY BLAZER boards a taxi headed for Heathrow Airport. A middle-aged consultant, James is on his way to a business meeting in Singapore. It is an unusually warm morning in London, with temperatures already around 22.2°C (72°F). The forecast calls for highs near 30°C (86°F), which is extreme for Londoners given average summertime daily highs of around 22.2°C to 23.8°C (72°F–75°F) and the fact that most of England averages at most a handful of days above 30°C per year.

Upon arriving in Changi Airport the next morning, James makes his way through a series of climate-controlled passages to the metro, which will take him to his hotel near city hall. The temperature in Singapore is a humid 32.2°C (90°F), which is about average for daytime highs in the month of June, and for that matter, pretty much every month in this consistently hot tropical climate where over 100 days hit temperatures above 30°C every year.

Thankfully for James, Singapore is among the most highly air-conditioned places on the planet. In fact, former Singaporean prime minister Lee Kwan Yew is famous for his penchant for climate control. He once remarked that air-conditioning was "one of the signal inventions of history," having "changed the nature of civilization by making development possible in the tropics." One of the first things Lee did upon becoming prime minister was to install air-conditioning in the offices where civil servants worked.[1]

Meanwhile, in the bustling streets of Bangkok, Thailand, an elderly woman named Chinda is starting her day in a fourth-floor walkup. Here too, the temperature is already over 32.2°C, with a heat index of an oppressive 40.6°C (105°F) given the humidity. The air in her cramped, un-air-conditioned apartment is stifling, but this is nothing out of the ordinary. She packs her belongings, dons a bandanna and hat to protect herself against the sun, and heads out to sell wares at a hawker stand in the central business district. Time is money for Chinda, and the busiest hour of the day is during lunchtime, when the heat happens to be at its most intense. Occasionally, she'll take a breather in the air-conditioned lobby of a nearby bank building, but she's hesitant to leave her stand unattended for too long.

Farther north, in rural India, a farmer named Raj is waking up to another day of intense heat. The monsoon is later than usual this year, and the scorching sun is desiccating his fields. Raj doesn't have much in the way of savings to fall back on, and he worries that his family will struggle to make ends meet this season. The combination of heat and drought have made it difficult for him to grow enough food to feed his family for two years running.

Raj's income is difficult to measure, as much of it consists of subsistence agriculture. It is likely less than the official average per capita income in India, which is approximately $2,300 per year, or roughly $8,300 adjusting for purchasing power. His story is all too common in rural areas across the developing world, where heat, drought, and food insecurity are a daily reality for millions of people.

The inequalities underlying the problem of climate change are likely familiar to you. Indeed, the dramatic contrasts outlined above may even feel a bit like overkill.

Given per capita incomes of $77,000 in Singapore ($127,000 PPP-adjusted), James enjoys the luxury of air-conditioned environments and quick transportation options that shield him from the worst of the heat. Air-conditioning may be less common in James's home country of England, but this is not for lack of resources, but a historical lack of need given the relatively cool climate.

In Thailand, where per capita income is around $7,000 ($18,000 PPP-adjusted), Chinda and millions like her have little choice but to endure the heat, with limited access to air-conditioning or other forms of relief, though incomes are fast approaching levels where rudimentary home air-conditioning adoption tends to become more widespread in places hot enough for year-round demand.[2]

Air-conditioning is certainly not a very realistic option for Raj, whose village has only intermittent access to electricity, whose family does not yet own a refrigerator, and whose income is likely barely $100 per month.

Of course, we already know that residents of the high-income world are responsible for most greenhouse gas emissions historically, though their share of annual emissions has fallen over time. Per capita emissions in the United Kingdom, which is on the low end of Organisation for Economic Co-operation and Development (OECD) countries, nevertheless stood at 5.2 tons of CO_2 per year in 2021, which was nearly triple that of India.[3] The average American emits around fifteen tons per year. The average Russian, South Korean, and Canadian, 12, 12, and 14 tons per year, respectively.

We know that these and other physical effects of climate change are likely to be more difficult to deal with for people like Chinda, who live in poorer countries, which are also hotter at baseline.

We know that climate change is likely to hit especially hard for those like Raj whose livelihoods depend on agriculture, given the adverse consequences of heat and drought for agricultural yield.

Each of these aspects of global climate inequality—the inequality in historical emissions, the inequality in incomes between rich and poor countries, and the inequality in vulnerability between rural and urban areas—is inseparable from the climate change problem, as we will explore in further detail.

At the same time, it is also true that many conversations around climate inequality and climate justice may employ an overly simplified heuristic, one that adheres to a familiar narrative of "guilty Global North, aggrieved Global South."

While this heuristic captures some of the most important dimensions of the problem of climate justice, at least those that have been most important historically, it may be also limited in critical, decision-relevant ways.

The next few chapters ask us to consider whether there may be other dimensions of climate inequality that are obscured by this familiar framing, and how it might affect our understanding of what can be done about the various injustices created by a warming world.

8

Climate Change and Economic Opportunity

MY PARENTS GREW UP on the outskirts of Busan, a port city on the southern coast of the Korean peninsula.

I can recall my father's stories about coming back from summer vacation to see desks emptied of elementary school classmates who had passed away over break—sometimes from preventable diseases like cholera, which festered in open sewers. Many Korean cities at the time lacked even the most rudimentary healthcare facilities, and health insurance was barely nascent.[1] Now, my mother and he occasionally fly back to Seoul from the United States to, among other things, receive state-of-the-art medical care. On the way, they FaceTime us over the free public Wi-Fi on the high-speed rail, which, at least in terms of average speeds, clocks more than double the Amtrak Acela in the United States.[2]

When my parents were born, average incomes in South Korea were less than $1,000 per year.[3] According to available evidence, this is a material standard of living that is less than what prevailed in countries like Italy and Spain during the fifteenth and sixteenth centuries, nearly 500 years ago.[4] Today, South Korea's per capita income—at $43,000 adjusting for PPP—surpasses that of Italy, Spain, and Greece, and rivals that of France and the United Kingdom. This represents an unprecedented rise in living standards—a more than twenty-fold increase within my parents' lifetimes.[5]

At the same time, my mother and father are not alone in observing that, increasingly, the economic fortunes of South Koreans appear to be bifurcating. While it is impossible to fully disentangle fact from nostalgia when their generation speaks of "being poor, but being poor together," their remarks appear broadly consistent with the data. Some, like tech mogul and Kakao founder Kim Beom-su or the K-Pop bands Black Pink and BTS, have benefited massively from recent economic developments, amassing billionaire fortunes in what seems like the blink of an eye. According to *Forbes*, there were at least forty South Korean billionaires as of 2022. At the same time, many Koreans appear to be increasingly left out of the accelerating knowledge economy, giving the country one of the highest poverty rates in the OECD, particularly among older generations.[6]

South Korea's economic experience may be unique in its rapidity and intensity. Few countries have seen such material gains in the span of a single generation.

But the twin dynamics of economic growth amid persistent and often rising inequality appear to be representative of broader forces at play throughout much of the world.

Growth and Inequality in the Twentieth Century

The past fifty years have been a remarkable period when it comes to global economic growth. Humanity has seen extraordinary increases in standards of living, averaging income growth of 4 percent per year between 1960 and 2015. During that time, per capita income for the average human being increased from $4,100 to $13,500, representing a doubling time of roughly eighteen years.[7] This is light speed compared to average annual growth in the nineteenth century of less than 1 percent per year, and as far as we can tell based on limited evidence, faster than all centuries prior.[8]

It isn't simply that we all have more physical accoutrements. This period has also ushered in remarkable gains on many dimensions of human flourishing. Global average life expectancy has increased from approximately 50 years in 1960 to over 72 years today.[9] In China, life expectancy rose from less than 45 in 1950 to over 75 in just over half a century. Global average literacy rates have more than doubled, rising from 42 percent to 86 percent over a similar period, a truly remarkable improvement in educational achievement. Far fewer mothers die during childbirth now than in centuries past; fewer people on average commit suicide; and, despite what news coverage may make it seem, violent conflict appears to have declined as well.[10]

We have amassed a tremendous amount of material wealth globally. And yet, it is something of a puzzle why we have not yet managed to eradicate material poverty completely. Economic thinkers of previous centuries may have been perplexed by this fact. John Maynard Keynes famously worried about the problem of too much wealth and not enough work in a fast-industrializing world. Or, as economic historian Bradford DeLong puts it, despite our technical ability to "bake a sufficiently large pie to create a veritable utopia wherein everyone feels safe and secure and is adequately fed," we have not managed to "slice and taste" the pie in such a way that achieves this for all of humanity.[11]

The diverging fortunes of those countries at—or catching up to—the technological frontier and those lagging behind remain stark. Some of the so-called tiger economies of China, Taiwan, and South Korea averaged GDP growth rates of up to 8 percent per year during the latter half of the twentieth century. In contrast, some African, Latin American, and Middle Eastern countries have experienced decades of stagnation, including prolonged periods of *negative* growth. The Democratic Republic of the Congo experienced average economic growth of −10 percent *per year* during the decade ending in 2000.[12]

In many tropical countries material standards of living remain at or below what Americans and Brits enjoyed at the turn of the twentieth century, nearly 125 years ago. Ghana's per capita income today is around $7,500 even after adjusting for differences in local prices.[13] And yet, Ghana is still among the richest third of African nations by income per capita.

In the context of climate change, it is important to note that many of the countries that have grown slowest and remain most mired in poverty tend to be closer to the equator, where the climate is already hotter to begin with. They are also, by and large, the countries that have contributed least to the stock of greenhouse gas emissions. Even though countries like Thailand and India have experienced rapid growth in recent decades, their standards of living remain at levels well below that of most developed countries like the United States, the United Kingdom, Germany, or Japan, all of whom have contributed more in per capita terms.

Another important feature of recent economic growth has been the persistence of (and in many cases, sharp rises in) inequality *within* countries—that is, across individuals, especially between the rich and poor within each country.[14] One might argue that this more local dimension of inequality is one that tends to be more viscerally felt. In the United States, while average earnings for workers with at least a bachelor's degree nearly doubled between 1965 and 2012, those without an undergraduate degree saw their incomes rise by less than 25 percent. For men without a high school diploma, real wages appear to have actually declined over the same period.[15]

The phenomenon of bifurcating labor market outcomes within countries is by no means limited to the United States, or the example of South Korea. The United States has long been one of the most unequal countries within the OECD, and it also saw one of the largest increases in inequality over the past several decades.[16] But similar trends in direction, if not in magnitude, have been documented in many other developed countries. Germany, New Zealand, Canada, Italy, and the United Kingdom all have seen significant increases in the Gini index—a measure of income inequality within a country—over this period.[17] As mentioned at the outset of the book, China has also experienced rapid increases in interpersonal inequality over the past several decades. In India, the share of income going to the top decile grew from 30 percent in the 1980s to 56 percent in 2020.[18]

What is causing such economic polarization remains hotly debated. One common feature across many countries is the role of technological change and how it affects workers with differing access to formal education and training. A growing body of scholarship observes that changes in technology appear to be playing a significant role in determining the

CLIMATE CHANGE AND ECONOMIC OPPORTUNITY   157

kinds of jobs that are in high and low demand. One estimate by economists Daron Acemoglu and Pascual Restrepo suggests that more than 50 percent of the changes in wage structure in the United States may be attributable to automation and technological change of various forms. As economist David Autor puts it, "almost every country has had to deal with some version of this phenomenon. It's just that some have been better at pushing against it than others."[19]

A perhaps concerning juxtaposition is what appears to be lower rates of economic mobility than many might expect, which is related to but distinct from economic inequality. Inequality is a measure of dispersion. The gap between the richest and the poorest 10 percent of the population, for instance. Mobility is a measure of the rate of churn between income groups. For instance, the likelihood that someone born into the bottom 10 percent will reach the top 10 percent in adulthood. If inequality is the number of floors in a high-rise, mobility could be thought of as the speed and accessibility of the elevators.

Some might consider upward economic mobility to be a defining feature of the American dream. When economists Raj Chetty, Nathan Hendren, Patrick Kline, and Emmanuel Saez measured the likelihood of upward mobility across 40 million Americans using anonymous individual tax records from the IRS, they found that such rags-to-riches dreams may in many places be elusive. The probability that a child born to parents in the bottom fifth of the US income distribution reaches the top fifth by adulthood averages around 5.5 percent in Gaston County, North Carolina, which is just outside of Charlotte. This compares to 18 percent in San Jose, California, and 19 percent in Boston, Massachusetts.[20] Notably, Chetty and colleagues found significant variation across races, with Black men facing some of the lowest rates of upward mobility, and some of the highest rates of downward mobility as well.

A striking feature of this research is just how much upward mobility varies at the highly local level. Even within the same city, rates of upward economic mobility can vary enormously. Take Boston, for instance. In a neighborhood like Back Bay, children born to low-income (bottom 20 percent of the US income distribution) families have a roughly 42 percent chance of reaching high income (top 20 percent) status by adulthood. In contrast, those born to similarly low-income parents in Jamaica Plain, less than four miles away, have a less than 1 percent chance of doing so.

In sum, the past half century has seen both unprecedented economic growth and significant economic bifurcation, both across countries as well as within. It is amid the backdrop of these twin dynamics that the potential economic and social implications of climate change must be examined.

Hazard, Exposure, and Vulnerability

It is worth introducing some vocabulary before proceeding further. A simple framework decomposes the adverse impacts of climate change, or climate risk, into three components: *hazard, exposure,* and *vulnerability.*

Hazard refers to the type of physical threat that climate change gives rise to. This could be extreme heat, increased rainfall variability, increased wildfire risk, sea level rise, or something else. Depending on the time and place, one set of hazards may be more locally important than another. Wildfire is a more acute hazard in places like Portugal and Spain within Europe, or California and the Pacific Northwest within the United States, at least with respect to direct fire-related damage. The relevant hazards will obviously vary depending on the time of year.

Exposure, broadly speaking, can be thought of as how much of a hazard is experienced by a given area or individual. Coastal residents will usually have greater exposure to flood hazards than those living inland. Households in tropical countries are on average likely to experience greater heat exposure than those in higher latitudes, though this will depend on other factors, like income or infrastructure. As we saw in the example above, residents of Singapore and Bangkok may experience on average very different levels of heat exposure, even though average climates and ambient temperatures are not so different. Exposure is, as we will see, often a function of both geography and human influences, including social and economic systems.

Vulnerability is typically defined as the propensity or predisposition to be adversely affected by climate change. It can apply to humans but also to natural systems, but for the purposes of this chapter we will focus primarily on the former. At the highest level, it refers to the extent to which such physical risks are likely to translate into adverse human outcomes.

Vulnerability can be thought of as the function that relates a given physical climate hazard to realized human suffering. As we will see, this is a key piece of the puzzle when it comes to thinking about how climate change relates to economic inequality.

The three concepts do not always form a neat partition. For instance, realized exposure can be inclusive of factors that determine vulnerability. Lower-income workers may be more vulnerable to wildfire smoke because they are more likely to have preexisting health conditions, but their exposure to it may also be higher because they are more likely to work outdoors or less able to bargain for flexible work hours.

Vulnerability is also closely related to the concept of *resilience*, which one might think of as the ability to bounce back from climate shocks. Climate resilience is defined by the Intergovernmental Panel on Climate Change (IPCC) as "the capacity of social, economic and ecosystems to cope with a hazardous event or trend or disturbance, responding or reorganizing in ways that maintain their essential function, identity and structure while also maintaining the capacity for adaptation, learning and transformation."

Vulnerability and resilience are also intimately connected to the concept of *adaptation*, which typically refers to systems or investments that help reduce the adverse impacts of a given hazard. For example, installing air-conditioning or heat pumps may be an important adaptation to life in a hot, humid environment. Sea barriers or riparian buffer zones may be adaptations to flood risk. Social norms like the Spanish siesta, or workplace practices that limit the amount of time working in direct sunlight may also constitute adaptation.

If the distinction between some of these concepts seems blurry to you, you're probably not alone. There is still some debate regarding the precise definitions and how they intersect. Given that the problem of climate change does not adhere to disciplinary boundaries, some continued semantic ambiguity is probably to be expected. For instance, one distinction that has been made is that resilience primarily comprises ex post actions such as coping or rebuilding, whereas adaptation typically happens ex ante, but with most climate shocks comprising repeated hazards, the practical distinction can be hard to make out.

Yet at the highest level, a group of people may be more vulnerable to climate change because they do not have investments in place to reduce the adverse impact of climate shocks (less adapted) and because they are, for whatever reason, less able than others to bounce back from a given shock (less resilient).

In this sense, the determinants of vulnerability, adaptation, and resilience are critical when trying to determine who is hurt most by climate change and why. They will also be increasingly important as societies begin to more proactively assess what to do about it, in devising policy interventions aimed at helping the poor adapt to and cope with a changing climate.

Geography and Exposure to Climate Hazards

Averages can mask important variation. Life expectancy at age 40 for American men was 79.4 years over the period from 2001 to 2014. This masks tremendous variation across individuals by income levels. For men in the top 1 percent of the US income distribution over that time, life expectancy was 87.3 years. For men in the bottom 1 percent of income, it was 72.7 years, roughly 14.6 years less.[21] This means that the difference in life expectancy between the poorest and richest Americans is similar to the difference in averages between the United States and Kenya.[22]

Analogously, professional basketball players tend to be very tall, with an average height of 6 feet, 6 inches among players in the NBA in 2021–2022. But whereas some players tower as high as 7 feet, 3 inches, there were also at least twenty players who were six feet or under.[23]

Thinking about the variation inherent to average indices is quite important when it comes to the physical dimensions of exposure to climate hazards. Talking about global average temperature increases, whether it is 1.5°C, 2°C, or 3°C, can be a useful heuristic. But there are

CLIMATE CHANGE AND ECONOMIC OPPORTUNITY 161

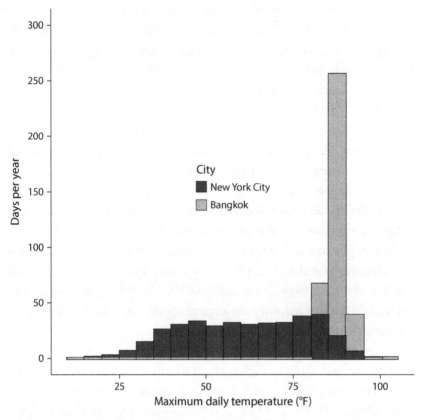

FIGURE 8.1. The distribution of maximum daily temperatures in a year over the period from 1990 to 2009 for New York City and Bangkok. *Sources*: Historical data are from ERA5 (produced by the Copernicus Climate Change Service [C3S]) and projection data are from twenty-two CMIP5 models based on the RCP8.5 scenario. Calculations by author.

dimensions along which such averages do not adequately convey essential information about extremes when it comes to local exposure and the resulting harm.

Let us start with how average warming translates into hotter days, and how this conversion varies across geographies. Here, a visual may be helpful. Consider figure 8.1, which plots the distribution of daily maximum temperatures for New York City. It shows how many days there are in a given year with highs in the 0°F–5°F range, the 5°F–15°F range, all the way up to days with highs of 100°F or above.

Overlaid on top is a similar figure, but for Bangkok, Thailand.

Note first that Bangkok is a much hotter place than New York on average. If we collapsed all the days into a single average, the average temperature in New York would be somewhere around 12.8°C (55°F) and in Bangkok around 28.9°C (84°F).[24] That is a difference of approximately 16°C (29°F).

It is worth noting that, even in the most extreme warming scenarios, the expected *change* in climate is likely far less severe than the existing *difference* in average climates between temperate and tropical countries. It is worth noting here that average humidity in Thailand is higher than in New York City. So each hot day in terms of absolute, so-called dry-bulb temperature is likely to be nontrivially less comfortable in Bangkok.

Another feature that jumps out is just how much more mass there is at the high end for Bangkok. One can see that, whereas a relatively small fraction of days in the year are above 26.6°C (80°F) or 32.2°C (90°F) in New York, the vast majority are already above 26.6°C in Bangkok, with a significant mass above 32.2°C as well.

In figures 8.2 and 8.3 I have plotted the historical daily temperature distribution and what the daily temperature distribution might look like toward the end of the century, given business-as-usual warming trajectories. Notice that the entire distribution shifts to the right in both cases. In New York City there are significantly fewer very cold days, but also more very hot days (above 90°F). In Bangkok there were not many cold days to begin with, so the relatively mild days in the 21.1°C–26.6°C range (70°F–80°F) have reduced in number, many shifting over into hotter days in the 26.6°C–32.2°C range.

What I'd like to draw our attention to is the uptick in the number of *very hot* days. These are days above the threshold of 32.2°C (90°F), a threshold I have chosen primarily to correspond with existing studies that tend to report temperatures in degrees Fahrenheit.

Figure 8.4 depicts the same shift but expresses it in terms of the change in number of > 32°C days for a given amount of average annual temperature increase.

The same increase in average annual temperatures translates into a much larger increase in very hot days in Bangkok compared to New York City.

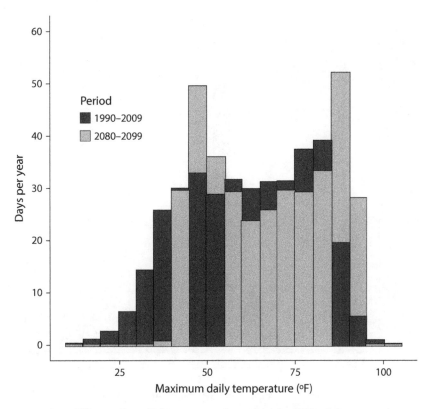

FIGURE 8.2. The number of days per year by maximum daily temperature in New York City, 1990–2009, compared to 2080–2099 using historical weather data and climate model projections. Days with temperatures above 80°F are lightly shaded and temperatures above 90°F are heavily shaded. *Sources*: Historical data are from ERA5 and projection data are from twenty-two CMIP5 models based on the RCP8.5 scenario. Calculations by author.

With some exceptions, this appears to be true as a general phenomenon as one gets closer to the equator. For a given amount of *global* warming, already hot, generally lower latitude localities will see a much larger absolute increase in the number of damagingly hot days. These places, for reasons noted before, tend to be poorer on average.

Of course, local climate is a complex phenomenon, and so this is not always the case. Whether lower latitudes correspond to greater increases in hotter temperature depends on a wide range of factors, including but not limited to position vis-à-vis the ocean; the direction of the ocean

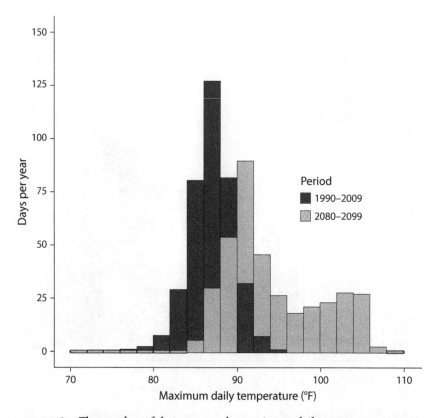

FIGURE 8.3. The number of days per year by maximum daily temperature in Bangkok, 1990–2009, compared to 2080–2099 using historical weather data and climate model projections. Days with temperatures above 80°F are lightly shaded and temperatures above 90°F are heavily shaded. *Sources*: Historical data are from ERA5 and projection data are from twenty-two CMIP5 models based on the RCP8.5 scenario. Calculations by author.

currents closest to the coastline, which determines how relatively cool or warm the water tends to be; whether trade winds blow inland or out toward the sea from on land; whether there are major mountain ranges nearby; and whether they cut north-south or east-west, among many others. For instance, some lower latitude regions of East Asia may see relatively limited increases in very hot days because what would otherwise be the hottest times of year coincide with the Indian monsoon (of course, the summer humidity may be dreadful). Conversely, some

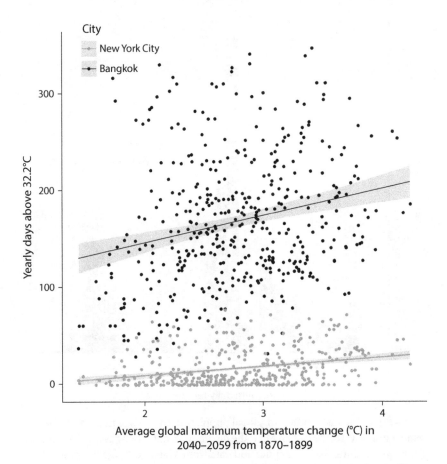

FIGURE 8.4. The relationship between average global temperature change (from 1870 to 1899) and the number of days above 90°F (32.2°C) for Bangkok and New York City. Each dot represents one model-year (between 2040 and 2059), and lines of best fit are calculated separately for each city. *Source*: The data are from twenty-two models from CMIP5 based on the RCP8.5 scenario. Calculations by author.

higher latitudes may experience sharper increases in heat because seasonal precipitation patterns typically feature dry summers and wet winters: places like Seattle, Washington, or Palo Alto, California, are cases in point.

While this phenomenon is moderated somewhat by the fact that higher latitudes will experience more rapid increases in average annual temperatures—that is, average warming will likely be larger in a place

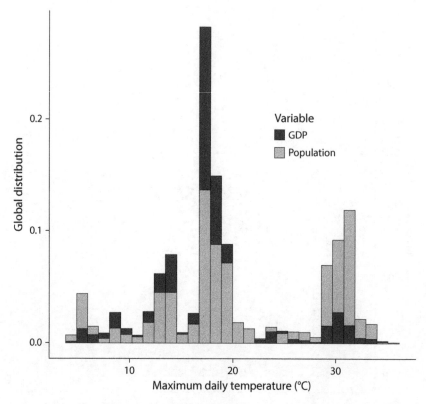

FIGURE 8.5. The distribution of global population by average maximum daily temperature (°C). Overlaid on top is the distribution of GDP by average maximum daily temperature. *Source*: Replication data from Marshall Burke, Solomon M. Hsiang, and Edward Miguel, "Global Non-Linear Effect of Temperature on Economic Production," *Nature* 527, no. 7577 (2015): 235–239.

like Anchorage, Alaska, compared to a place like Dhaka, Bangladesh—it is typically not enough to offset the fact that a given mean-shift in global average temperature will lead to a much sharper increase in the number of very hot days for already hotter and poorer places.

Figure 8.5, adapted from a study by Sol Hsiang, Marshall Burke, and Ted Miguel, illustrates this point nicely. There are many more people living in the tropics than in higher latitudes, and many more people with incomes below $10,000 per capita than richer people with incomes above it.

In short, geography is the first factor that makes climate damage likely to be larger for poorer people. Climate change will lead to much larger increases in heat for people in already hotter climates. This is probably not news to many readers.

Whether and to what extent people in poorer countries will be able to adapt by moving to less exposed places is an important question that, as discussed previously, may be influenced by many factors, some of which depend on policy choices made today.

For now, we can surmise that, unless poorer people are somehow better adapted to hotter temperatures, whether physiologically, culturally, or institutionally, and this adaptation more than offsets the handicaps of being poor, the *marginal* impacts of a given hot day may also be larger in poorer countries.

Even if this is not the case, the sheer magnitude of exposure in most poor countries suggests that the adverse consequences of heat discussed in the previous chapter will stack up disproportionately for developing countries. Whether it is the effects of heat on human health, labor productivity, learning and human capital formation, or something else, the *aggregate* impacts in terms of human suffering will likely be much larger for poorer countries due simply to the fact that the poor will face so much more of it.

It is worth noting here that the damages in terms of easily monetized financial losses may be less concentrated in the developing world. This is primarily because the market damages associated with climate change will occur on a smaller base of financial assets in poorer countries. Suppose there are two neighborhoods with an equal number of households: one where all homes are valued at $1 million and another where they are valued at $100,000. Flooding 5 percent of homes in the first neighborhood may have a similar monetized impact on real estate as flooding 50 percent of homes in the second, even though the realized human suffering will arguably be much more extensive on the latter than on the former. Similarly, if hotter days reduce labor productivity equally across the world, the monetized impact may be lower in the Global South because of lower baseline wages, even if the relative productivity hit is larger in percentage terms.

As we have already seen, many if not most of the damages from climate change may come in the form of nonmarket assets, like health, human capital, and conflict, which tend to be harder to measure. From the standpoint of understanding the extent of climate inequality, accurately measuring the nonmarket damages may be critical, challenging as this may be. Even when they are measured, policymakers may want to give more weight to the same proportional damage suffered by poorer people based on equity and other factors. This is a question of *equity-weighting*, which remains a controversial issue in regulatory analysis, and which we will revisit in chapter 12.

It is also worth paying attention to both tails of the figures above. A given amount of warming results in an increase in very hot days for both New York City and Bangkok, but for New York City also a reduction in very cold days. This means that, on net, the damages associated with this warming may stack up differently in these two places because of the relationship between cold days and various dimensions of human well-being. For instance, warming that takes the form of a reduction in heavy frost days is largely beneficial for agricultural yields. Warming that takes the form of additional hot days can be highly damaging. In the literature, days with high temperatures above 29°C (84.2°F) are referred to as "killing degree days," in reference to the fact that maize yields decline sharply when crops experience such hotter days.

While everyone will experience more heat, in northern latitudes there will also be a sharp reduction in cold days. The evidence from part II, where we discussed the temperature-mortality relationship, suggests that this is important in thinking through climate damage inequality. Just as we now know that there is a strong causal relationship between hotter temperature and all-cause mortality, there is a strong causal relationship between colder temperature and higher mortality as well.

This means that the net health effect of climate change may be attenuated—and possibly even reversed to be positive—in some richer, colder countries. In the United States we have evidence to believe that the net impact may still be to increase mortality, because extreme heat appears to be more nonlinearly damaging to human health than extreme cold for areas of the country that are poised to see the largest increases in heat.[25] But this does not undo the fact that, in comparison to a place like Bangkok, the way global warming manifests for New Yorkers will have some pluses that offset at least some portion of the (likely already smaller) minuses.

This contrast becomes clearer if we switch continents and consider the effects of temperature on health for residents of Berlin, Germany. Here, because climate change will lead to a much sharper reduction in cold days than it will increase hot ones, the net projected mortality impact as estimated by the Climate Impact Lab is a reduction of 150 deaths per 100,000, or a roughly 15 percent decrease. This makes sense given that colder temperatures kill far more people in places like Germany, Norway, and Canada. As mentioned before, the exact pathways through which colder temperatures increase mortality are yet unclear, but leading suspects include compromised immune systems, greater spread of infectious diseases (due in part to more indoor activity), and the fact that cold increases blood pressure and makes blood thicker, leading to elevated risks of cardiovascular and other diseases.

Compare this to the effects of heat on health in a place like Accra, Ghana, which can expect to see roughly 120 additional days above 32°C (90°F) per year. The Climate Impact Lab researchers project that this warming will lead to an increase in the death rate of 160 per 100,000, which is a roughly 14 percent increase in all-cause mortality. This is even after accounting for the fact that Ghanaians are likely better acclimated to heat than those in higher latitudes at any given income, and for the expected distribution of future income growth rates in Ghana. Without adjusting for adaptation, the projected increase is 260 per 100,000, nearly twice as much.

Residential Sorting

Now I'd like to draw our attention to a similar but perhaps subtler dimension of climate inequality. One that is far more local, that manifests across cities within the same country, and even across neighborhoods within the same city. It is also one that is deeply intertwined with—perhaps exacerbated by—the way markets tend to work.

Across the United States, hotter parts of the country generally tend to have lower average incomes. This isn't to say that there aren't many high-income individuals living in the South, or that there aren't people in poverty in northern states like Massachusetts or Washington. But rather that, on average, incomes in the South and Southwest tend to be lower than in the Northeast or Pacific Northwest. This correlation is, perhaps surprisingly, true in other countries, including much of North, Central, and South America. Across many countries including the United States, Mexico, Peru, Bolivia, Brazil, Panama, and Venezuela, municipalities that are 1°C (1.6°F) hotter on average have 1.2 to 1.9 percent lower per-capita incomes.[26]

Often, such relationships hold at far more local levels. In California, individuals in the lowest quintile of income live in zip codes that experience 70 days per year above 90°F (32.2°C). For those in the highest quintile, the corresponding number is 26 days per year. In other words, poorer Californians experience nearly three times as many days per year with potentially dangerous heat than richer Californians do.

This difference is not simply because richer people tend to live in cities near the coast, like San Francisco and Monterey. Such temperature-income gradients appear to hold even within cities as well.

Figure 8.6 shows two maps of Los Angeles. The one on the left depicts average relative surface temperatures and the one on the right depicts average residential income by census tract. These within-city differences can be large. In my own amblings around Los Angeles, I have often been amazed at how, on the same day, the temperature in coastal and tree-lined Santa Monica can be in the low 80s°F (26–29°C), while the temperature in Downey or somewhere in the Inland Empire can be over 100°F (37.7°C). Such systematic differences in local heat have been documented in dozens of other cities as well.

This begs the question of why local climates tend to be so correlated with income. One explanation is that it is the result of market forces. If extreme heat is generally undesirable, then would-be homeowners and renters would likely be willing to pay a premium for less extreme and thus more desirable locations, all else being equal. In other words, the dynamics of supply and demand in the housing market may naturally generate a premium for housing in areas like Santa Monica where the local climate is more desirable relative to places like Downey. Other factors may layer on top of this. If space in desirable locations is limited due to geography, or perhaps constrained due to local ordinances (e.g., restrictions on density), demand may further exceed supply, bidding up prices in these desirable areas even more. If traffic is congested and commute times swell with distance, such premiums might be even higher, since the cost of getting from point A to point B increases with congestion.

Of course, all other things are rarely equal: Santa Monica may be desirable not only because of its local climate but also because of its access to hip nightlife and proximity to tech jobs. And the homes in Santa Monica may on average be bigger and newer and fancier than in Downey. But to the extent that one could make "all else equal" comparisons, economic theory would predict an "amenity premium" associated with milder climate.

This prediction appears to be borne out in the data. According to analyses by economists David Albouy, Walter Graf, Ryan Kellogg, and Hendrik Wolff, and similar analyses by Paramita Sinha, Martha Caulkins, and Maureen Cropper, Americans systematically favor places that have fewer extreme days, and may be willing to pay a premium for it. In one study, Albouy and colleagues tested this hypothesis using data from over 2,000 American cities. Using detailed daily weather data and a wide range of housing and city characteristics (e.g., number of bedrooms, access to jobs), they tried to come as close as possible to such an "all else equal" comparison. The conclusion was that American households are willing to pay a significant premium to live in places with more pleasant climates. People are willing to pay to avoid both very hot and very cold days, and they prefer days with highs in the mid-60s (°F) (15–20°C) if at all possible.[27]

Importantly, these estimates control for the fact that more extreme climates will require greater heating and cooling expenses. What they are picking up here is what one might call the *intangible* value of living in a more pleasant local microclimate: the value of being able to go on more walks outside with the kids or the dog, of not sweating profusely during the walk over to the local subway or bus station, or simply having to worry less about the potential power outage during a heat wave or a major winter storm. This means that, even within a given state or city,

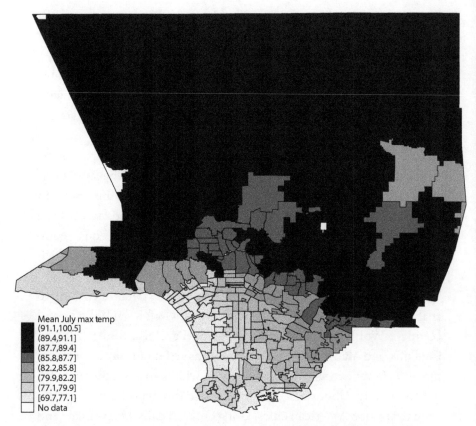

FIGURE 8.6. The map above shows average annual (or July) temperature (°F) from 2000 to 2015 in Los Angeles, by zip code. The data are from the PRISM model. The map on the following page shows mean 2018 income in Los Angeles, by zip code. *Source*: The data are from the IRS. Calculations by author.

the rich pay to live in places that are naturally sheltered from the most extreme climate shocks, and the poor often end up being priced out of the most pleasant local climates, and they also still must pay more in resulting heating and cooling expenses.

Such studies have been replicated in their broad contours in Italy, the United Kingdom, and other countries, with similar findings.[28] And within the United States, many cities appear to exhibit similar correlations between neighborhood-level income and the frequency of hotter days. It seems that, as a species, we pay to live in places that suit our temperature tolerances—at least when we can afford to. Not unlike

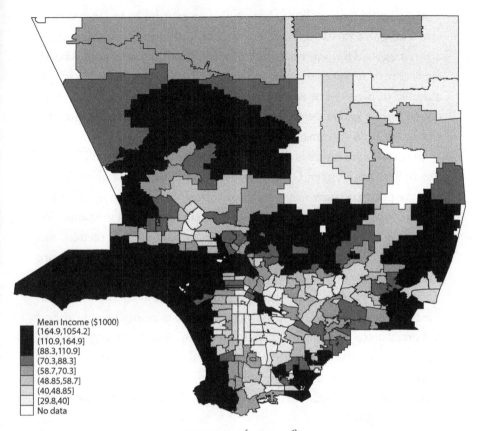

FIGURE 8.6. (*continued*)

Goldilocks and the Three Bears, we appear to have a systematic preference to have it not too hot, not to cold, just right. In fact, evidence suggests that in relatively cold countries, the relationship between temperature and income is reversed, whereby richer households tend to be located in warmer parts of the country, consistent with the Goldilocks hypothesis.[29]

Of course, there is some degree of variation in this preference for non-extreme climates. For instance, Albouy and coauthors found that the elderly have a stronger preference to avoid cold, which is consistent with the cold–mortality relationship mentioned earlier (which is driven primarily by those above 65 years of age) as well as perennial elderly complaints about what the cold does to their arthritis. As we saw in previous chapters, individuals' specific susceptibility to temperature stress may also vary by gender, body mass, and a number of other characteristics. And in some countries, other factors like proximity to coasts, elevation, or access to major transit networks may dominate, leading to a muted or reversed relationship between average temperature and income. But by and large, we seem to prefer to live and work in climates that are not too extreme.

In some cases, there is a racial overlay to this disparity. In US cities like Baltimore, Denver, Dallas, New York, and Miami, neighborhoods that are poor and have more residents of color tend to be significantly hotter in the summer than wealthier, whiter areas within the same city.

An important contributor appears to be the amount of greenery and tree cover. "Walking through different neighborhoods, there is a stark difference between places that have lots of greenery and places that don't," observes one Richmond, Virginia, resident in an interview with the *New York Times*, whose lower-income, high-minority neighborhood has greenspace covering just 12 percent of the surface area compared to 42 percent in predominantly white neighborhoods nearby.[30] Other sources of variation in local heat island effects include the presence of heat-creating industrial facilities, heat-absorbing asphalt parking lots, and large highways.

Some studies trace the origin of such differences to policies including redlining, whereby the federal government in the 1930s created maps that rated the riskiness of different neighborhoods for real estate investment. It is difficult to know what the causal effect of redlining was on neighborhood tree cover and other determinants of the urban heat island effect because the neighborhoods thus categorized are correlated with so many other factors—like local tax revenue and demographic composition—that could eventually bring about such differences. But we know that housing is an important mechanism for building wealth, and that redlining affected access to housing finance differentially across racial groups.

These racial differences have engendered calls for action. "We can see that racial equity and climate equity are inherently entwined, and we need to take that into account when we're building our capacity to prepare," notes the sustainability manager of one city. Indeed, some cities are putting racial equity at the core of their efforts to enhance climate resiliency.[31]

Such efforts give us a window into the complexities of policy design, and arguably the need for better data to make informed adaptation decisions. These efforts often include major park-building and urban renewal campaigns, which can have many climate-resiliency benefits. More trees and urban park space can reduce heat and local air pollution. They can even confer flood protection benefits by holding more rainwater in root systems, reducing the severity of flash floods.

But such place-based policies can sometimes carry unintended consequences that may blunt or even reverse equity benefits. Because urban renewal projects tend to make neighborhoods more attractive, real estate values will tend to rise as demand for homes in the area rises. Unless the minority residents such policies were intended to target are already homeowners (which is less likely given the strong links between income, race, and homeownership), they will be unlikely to capitalize such gains financially. If anything, renters may be more likely to be priced out of their residences, and well-intentioned climate resiliency initiatives may end up accelerating local gentrification.

This does not mean that such policies are necessarily ill-advised, just that the problem is complicated by underlying market forces. This example illustrates two important problems. The first is that, when it comes to the racial dimension of climate justice (or environmental justice broadly), it is often difficult to disentangle problems of racism per se from the problem of poverty, or some other factor leading to differences in realized climate exposure. The second is that, for policies motivated by racial justice to have their intended effect, it is important to have a clearer picture of the causes that give rise to such inequities, and how the policies will interact with these and other economic forces.

It is possible for well-intentioned policies to, without an accompanying diagnosis of the underlying drivers of inequality in climate damages, end up expending public resources in a way that does little to address the inequities they originally intended to alleviate. One can imagine alternative policy designs that take such forces into account—for instance, by targeting disadvantaged households as opposed to particular places, or coupling place-based policies with other supports. While the information inputs necessary for enacting such dovetailed policies are nontrivial, they may be increasingly within reach, as I will argue in chapter 12.

To reiterate, an important piece of the puzzle is how markets respond. If poorer people generally face harsher trade-offs in life and thus view things like environmental quality as luxuries they can ill afford, the dynamics of the housing market may push them out of improved neighborhoods back into environmentally deficient but cheaper neighborhoods.

On the other hand, if poorer people would prefer to spend more of their own income on local public goods such as parks, greenspace, or air-conditioned classrooms, but find it harder to mobilize collective political voice, then solving such collective action problems through broader policy measures may be desirable from the perspective of the public good.

As we will see in chapter 9, such dynamics may play out in ways that sift poorer people into many climate-exposed jobs as well.

9

Climate Inequality
Close to Home

WHETHER ONE ENDS UP living in tree-lined Santa Monica, California, or more concrete- and asphalt-dominated Downey, California, is not simply a matter of willingness to pay. It is also a matter of ability to pay. Here the broader drivers of inequality become important to consider, especially as we look forward to the climate change that is in store.

A comprehensive account is beyond the scope of this book, and there is far from consensus in the literature. But it may help to at least have a feel of some hypotheses and the evidentiary basis behind them. Similar to how generating a prognosis for a patient can benefit from a better understanding of preexisting medical conditions, it may be useful to get a sense, even an approximate one, of how the onset of climate change may interact with preexisting societal conditions. So let us build some more intuition regarding the contours of labor market inequality in the modern economy.

Take the following four jobs at Amazon, which can be thought of as loosely representative of important trends in the modern labor market. Much of the following comes from my own perusal of job ads on sites like Indeed.com.

Amazon job number one is a senior technical program manager. The advertised salary range is $145,000 to $206,000 per year. Minimum requirements include a bachelor's degree in engineering, computer science, or a related technical discipline. Also required are "excellent leadership skills," "hands-on project management skills," and an "ability to communicate clearly and compellingly at all levels of the company." These are the *minimum* requirements. *Preferred* qualifications include "an advanced technical degree" (e.g., a master's degree or a PhD) "or an MBA." This is also a senior position, mainly for those who already had their foot in the professional door several years prior.

Job number one is a job reserved only for the unicorns of the knowledge industry—people with deep technical expertise, strong emotional self-regulation, and strong interpersonal managerial capacity: the trifecta of hard, soft, and social skills. This is in addition to all the basics, like writing proficiency and time-management and professional etiquette. The headline salary puts this job in the top ninety-fifth percentile of the US individual salary income distribution,[1] and this salary of course does not capture benefits, which most likely include performance bonuses in the form of stock options, generous health insurance and pension contributions, and the option of working in an office brimming with all of the hip amenities we have come to associate with high-end corporate jobs—baristas, yoga instructors, free organic snacks, and so on.

Job number two is more representative of someone in the middle to upper-third of the US personal income distribution. To be a business development representative at Amazon requires at least a bachelor's degree, but nowhere near as much specialized knowledge or experience as the first position. Pay ranges from $47,000 to $72,000, possibly with some reduced stock options, similar health insurance and pension offerings, and a similar set of on-site amenities.

Many more Americans would be qualified for this job. Still not the majority: of a population of roughly 300 million, somewhere on the order of 200 million are in the workforce with the rest being either too young, too old, or otherwise incapacitated. Of this workforce, less than half have completed a bachelor's degree. Of course, not everyone with a bachelor's degree who applies will get the job. But at least they would have a shot.

This leaves the other 100 million-plus workers. What are some jobs at Amazon that they might be eligible for?

Job number three, which Amazon seemed eager to fill, was that of delivery station warehouse associate. Here, the way pay is described may itself be an indication of the underlying differences in job stability. Instead of an annual salary range, we are provided an hourly rate, in this case, "up to $17.95 per hour."

Perhaps it was a function of Amazon's breakneck expansion at the time of this inquiry, but the requirements were minimal: "Immediate openings available. No resume or previous work experience required." Still, some requirements were listed, explicitly including the ability to "read and speak English for safety"; "lift up to 49 pounds"; "stand, walk, push, pull, squat, and reach during shifts"; and "work at a height of up to 40 feet on a mezzanine floor." Some indication as to potentially unpleasant or hazardous working conditions were provided up front: "You'll be working around moving machines—order pickers, stand-up forklifts, [and] turret trucks," and "Even with climate controls, temperature can vary between 60°F [15.6°C] and 90°F [32.2°C] in some parts of the warehouse; on hot days, temperature can be over 90°F in the truck or inside trailers."

At forty hours a week, the annual salary equivalent would be in the range of $35,000, assuming the high end of the pay scale for this position. This job would land somewhere between the bottom and second-to-bottom quintile of the US individual (pre-tax) income distribution. If employed full-time, workers would likely receive health insurance and workers' compensation for injuries on the job.

Finally, there is a large swath of workers, mainly in the bottom quintile of the income distribution, for whom jobs are often unlisted on such sites. In this fourth bucket are positions like janitor, refuse collector, landscaping worker, dishwasher, or food preparation worker. Many of these jobs pay less than $30,000 per year and are often advertised by word-of-mouth and other informal means or outsourced completely to companies specializing in such support services. Amazon, like any other large company, needs crews to clean and maintain their facilities— whether at flagship corporate campuses or the many distribution centers worldwide. While it is possible that working as a janitor or a landscaper at a place like Amazon may pay a premium relative to similar roles at smaller companies, the available data suggests that it is unlikely for such workers to make more than $40,000 per year.

This simple exercise helps to illustrate some broader trends. First, it exemplifies how specialized the modern economy has become. Specialization and the gains from trade are in part what have enabled the kind of unprecedented growth in material well-being we have seen in the past fifty years. Such specialization appears to have been further abetted by the tide of technological change (e.g., the IT revolution or industrial robotics), which can in principle lift all (or most) boats by freeing up human labor for more productive activities.

But specialization also entails the potential for greater inequality, particularly if technology makes certain tasks more valuable in the marketplace than others, and if the skills required to complete those tasks are costly to obtain. The marginal value of having slightly better delivery-management software is significantly higher when the pool of online shoppers is global. The marginal value of a slightly better janitor or landscaper or construction worker, perhaps not so much. Of course, even with such forces at play, widening inequality is not inevitable. Government policies, whether as part of the tax code, labor market policies like the minimum wage, or programs like health care or public education, which act as a form of nonmonetary transfer, can mitigate or exacerbate these inequalities to varying degrees—as can societal norms around pay disparities.

The second broader trend that this exercise underscores is just how important education and skills—both technical and interpersonal—are becoming for success in the modern economy. They may not be sufficient conditions for success, but they seem increasingly necessary to pass the first employment screen. Recent work suggests that, while having strong backgrounds in science, technology, engineering, and math (STEM) fields can lead to fast wage growth early on in one's career, those with "softer" skills—psychology, sociology, or English majors—can often catch up in earnings over time, especially as technology raises the returns to effective collaboration, suggesting that both "hard" and "soft" skills continue to be in high demand.[2]

Third, it underscores just how wide the gaps in compensation, and usually as consequence, ability to pay for protection against various environmental hazards or to care for one's health generally can be. These gaps are important in part because they lead to very different choice sets—whether in terms of housing location and quality, or the trade-offs involved in responding to various climatic stressors—be they mundane events like hotter days or more jarring ones like major hurricanes or wildfires.[3]

How did we arrive at such inequality? It is beyond the scope of this book to provide a comprehensive treatment of the many potential causes of these trends. It is often the case that observations about whether something is happening are easier to make than why. Everyone can observe an apple falling from a tree. Connecting it to a theory of gravity and generating convincing scientific evidence in support of that theory is another story. When bubonic plague (also known as the Black Death) hit Europe and Asia in the mid-1300s, it would have been difficult to ignore its carnage. Few at the time would have been able to trace the cause of the deaths to a bacterium spread by fleas. Similarly, it is harder to pinpoint the causes of such trends in economic inequality. Many hypotheses are proposed. Some refer to forces in the external environment. Others, to choices individuals make regarding how much education to pursue, or how much effort to exert at work.

One hypothesis that may be relevant for developed and developing countries alike involves the labor market impacts of technological change. In broad strokes, the idea is that technological change increases the productivity and market value of some kinds of skills more quickly than others, directly substituting for some while being highly complementary to others. A classic example is provided by the decline of jobs like bank tellers or manufacturing assembly line workers, where technology has tended to make some jobs more lucrative and others obsolete.

David Autor, Frank Levy, and Richard Murnane provide compelling evidence of this phenomenon. They categorize jobs according to their "routine-task intensity," where routine tasks are those that are repetitive and follow clearly defined rules and thus can be more easily automated. They show that workers in jobs that are more routine-task intensive have experienced relative declines in labor market prospects. Importantly, routine tasks are not limited to manual work, as in some blue-collar jobs like furniture manufacturing or food processing. Many jobs historically associated with white-collar occupations also appear to be routine-task intensive, including jobs that used to be conducted by bank tellers or travel agents.[4]

It is instructive to think about how much less labor-intensive the act of depositing a check into a bank account has become on average. Many such transactions used to be executed by hand, by bank tellers across thousands of brick-and-mortar bank branches. One would walk into a physical bank branch, and hand a check to a teller, who would then write down the associated account number and amount, examine the bill for authenticity, and begin the paperwork required to credit and debit the respective accounts party to the transaction. Of course, one can still do this at a local bank branch. But many checks are now deposited electronically, using an advanced ATM or an app on one's smart phone. Consistent with this idea, employment and wages for bank teller–like jobs have fallen in many countries, whereas job prospects for less routine-task intensive roles, for instance, the computer programmer who designed the app that handles online check deposits or the manager who coordinated the workflow between coders and designers and marketers who helped roll out the app, have risen rapidly.[5]

Of course, there are many other potential drivers of interpersonal inequality—including international trade, changes in market concentration, and changes in policies like the minimum wage or support for unions, to name a few. Yet across many countries the divergence in earnings power between those with and without access to higher education appears to be a consistent feature. The way those without the skills most relevant in a fast-evolving technological landscape have seen their economic fortunes decline in recent decades, and the fact that this phenomenon appears to hold across many different countries, suggests that these trends may be relevant for how we think about the distribution of climate damages as countries continue to industrialize.

In the fast-industrializing South Korea of the 1970s and 1980s, many of the sons and daughters of farmers and manufacturing line workers became bank tellers and travel sales agents. For the many families who converted those economic gains to further human capital accumulation, their sons and daughters may now be doctors, lawyers, and computer programmers. Whether those middle-link jobs will be as plentiful as low- and middle-income nations industrialize is an open question, which suggests that understanding how climate change and technical change interact in the labor market may be an increasingly important issue.

Occupational Sorting

Work is about more than simply earning a living. It is also for most people an important source of dignity, pride, and a sense of purpose. For many it can also be a critical determinant of physical health.

In the case of the hypothetical Amazon workers we examined above, some are more obviously exposed to climate hazards than others. It seems almost trivial to observe that warehouse and landscaping workers may face more extreme environmental conditions on the job compared to office workers. On hot, polluted days, we might expect employees in jobs one and two above to spend most of their time in climate-controlled offices, or perhaps working from home. Such accommodations seem less likely, though not impossible, for landscapers given where most of their work must occur. Whether air-conditioning is available for warehouse workers is an empirical question for which we would need better data, but anecdotal reports suggest that many large warehouses are not fully air-conditioned.[6]

If one accepts the premise that most work in modern societies is voluntary, this discussion may perhaps seem a bit paternalistic. Is not all employment "at will," meaning that workers could walk away if the conditions were unfavorable or unsafe, or if they deemed their compensation not to be worth the risk? In which case, on what basis do we concern ourselves with whether some workers are more exposed than others? We will discuss the important issues of choice, compensation (or hazard pay), and flexibility momentarily.

For now, let us examine table 9.1. It shows, for a handful of occupations in the United States, the amount of time spent either outdoors or indoors without climate control while on the job. Let's call this number

184 CHAPTER 9

TABLE 9.1. Average exposure, years of education, and employment for select occupations from 2007 to 2018. Calculations by author.

OCCSOC	Job title	Percent days exposed	Years of education	Employment (thousands)
53-4021	Railroad brake, signal, and switch operators	99.9	12.6	21
37-3011	Landscaping and groundskeeping workers	98.5	10.9	878
53-7081	Refuse and recyclable material collectors	98.5	11.4	120
47-2221	Structural iron and steel workers	98.4	12.1	65
47-2061	Construction laborers	95.8	11.1	895
43-5021	Couriers and messengers	75.5	12.9	82
47-2152	Plumbers, pipefitters, and steamfitters	73.3	12.2	393
35-9021	Dishwashers	64.5	10.6	506
47-4011	Construction and building inspectors	58.3	13.8	93
11-9021	Construction managers	40.6	13.9	226
41-3021	Insurance sales agents	35.8	14.5	353
19-3011	Economists	19.0	18.3	16
13-2051	Financial analysts	12.9	16.5	254
29-1011	Chiropractors	9.6	20.4	30
23-1012	Judicial law clerks	6.4	19.4	15
15-1134	Web developers	6.0	15.4	121
29-2034	Radiologic technologists	4.8	14.3	201

Source: Exposure data is from the US Department of Labor (O*NET); education data is from the American Community Survey; and employment data is from the Bureau of Labor Statistics.

the *occupational exposure score*. Specifically, it represents 100 minus the fraction of workdays that workers respond that they spend working in climate-controlled spaces. The data comes from O*NET, which surveys workers and occupational experts across 700-plus detailed occupational categories in the United States.

Not surprisingly, the average web designer and the average financial analyst spend most of their working hours in climate-controlled spaces. Also not surprisingly, workers in professions where work occurs outdoors have a very high exposure score. There are also jobs—building inspectors, for instance—that appear to involve a mix of time outdoors and indoors, with possibly varying degrees of climate control.

What might we expect the relationship between occupational climate hazards and income to look like? Two possibly competing forces are at play. To the extent that exposure to extreme environmental conditions represents a disamenity—something that we'd prefer to experience less of if possible—we might expect workers to demand higher pay to compensate for the added risk and discomfort. This is what economists refer to as *compensating differentials* or *combat pay*, and is not unlike the premium employers must pay to otherwise similar workers to work the night shift or to work weekends.

On the other hand, in a world where starting endowments are sufficiently different, poorer individuals may face harsher trade-offs in life and thus may be willing to take on overall less pleasant and riskier jobs even for lower pay. If access to education is hindered by credit constraints or other barriers (e.g., lack of information regarding the future job options that a college education may provide access to), we might expect such patterns—what economists call *utility dispersion*—to be further accentuated.[7]

Which force dominates is an empirical question that has not yet been fully answered. However, the available evidence suggests that, on net, climate hazards on the job (and in many though not all other contexts) may be disproportionately borne by lower-income individuals.

Figure 9.1 plots the average time a worker spends potentially exposed against average educational attainment, measured in years of schooling, for over 700 occupations in the United States in the year 2015. The size of the bubbles denotes the number of workers in each occupation. Clearly, exposure to climate hazards on the job is heavily concentrated in occupations typically chosen by individuals with lower levels of educational attainment.

Paul Stainier and I wanted to better understand which workers are more likely to be exposed to climate hazards on the job. Looking at data from the US Census, we observed that individuals without a bachelor's degree are over four times as likely to work in a highly exposed occupation, which we define as a job where the majority (over 60 percent) of

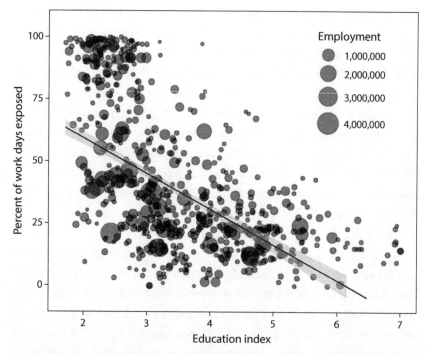

FIGURE 9.1. The relationship between average years of schooling and the percent of workdays exposed for 658 occupations in the United States. The percent of workdays exposed is defined as 100 minus the percent of workdays indoors in an air-conditioned environment. The line of best fit highlights the inverse relationship between education and occupational exposure. The sizes of the dots correspond to average employment in each occupation over the period from 2007 to 2018. *Source*: The data comes from the Bureau of Labor Statistics and the US Department of Labor (O*NET). Calculations by author.

workdays are spent potentially exposed to temperature extremes.[8] In fact, the bachelor's-degree-versus-no-degree discrepancy exceeds any differences across race or citizenship status. White Americans *without* a bachelor's degree are 3.5 times more likely to work in a highly exposed occupation than the average Black, Hispanic, or Indigenous worker *with* a bachelor's degree. Similarly, while non-US citizens are 1.5 times more likely than citizens to work in highly exposed occupations, this difference pales in comparison to the differences by educational attainment noted above.

Such differences in occupational exposure may be important determinants of the lived experience of climate change. After all, work is where most adults spend most of their waking hours. Wages from work also constitute the primary source of income for most people—both in the United States and globally.

Of potential concern as the world becomes hotter is the way heat may interact with other workplace hazards. For many readers, it may be difficult to intuit just how many incredibly dangerous jobs there are out there. In warehousing, roughly 5 out of 100 full-time-equivalent workers are seriously hurt on the job each year, one of the reasons why turnover in such jobs tends to be high. Injury rates are even higher in some industries, like correctional institutions (9.2 per year), animal slaughterhouses (7.4), or iron foundries (7.3). Some jobs that aren't typically associated with obvious health hazards also have high on-the-job injury rates, including couriers and express delivery services (7.5 per year) and automobile manufacturing (5.9). These figures are of course limited to incidents that are reported, in this case to the Bureau of Labor Statistics. Researchers suggest that many more, less serious incidents such as bruises and sprains, or slow-onset illnesses less obviously connected to work activities often go unreported.[9]

Though it is difficult to know how rates of unreported injuries compare, it is worth comparing these figures to injury rates in more white-collar occupations. In finance and accounting, for instance, official records suggest fewer than 0.8 out of every 100 full-time workers are hurt on the job each year.[10]

Nora Pankratz, Patrick Behrer, and I were interested in how these factors may interact. Given what heat can do to human physiology and cognition, it seemed plausible that hotter temperature and dangerous workplaces may be a combustible mix.

Examining claims data from over 15 million workers' compensation insurance claims for the state of California, we found that hotter temperature increases workplace injuries significantly. On a 90°F (32.2°C) day, average injuries increased by upward of 5 to 10 percent.[11] These echoed findings from previous work by economist Marcus Dillender looking at similar data in Texas, as well as for mining workers across the country. In some of the most heavily exposed industries, like mining, the effect of a day in the 90s can be upward of 50 percent.[12]

What is interesting here is the fact that this effect persisted in some indoor workplaces: in industries like warehousing, manufacturing, and wholesale. These are often settings where the lines between indoor and outdoor work can be blurry (e.g., an auto mechanic's garage) or where indoor spaces can be additionally costly to cool, perhaps because of sources of ambient heat (e.g., iron furnaces) or the sheer volume of space (e.g., large shipping warehouses).

Consistent with the notion of utility dispersion, it appears that lower-income workers are significantly more likely to be hurt on the job due to heat. The overall risk associated with hotter temperatures, at least in our data, is at least five times higher for workers in the bottom quintile of the income distribution relative to the top.

Why is this the case? One explanation could be individual preferences. Some people just tend to enjoy working outdoors or in jobs that require physical dexterity, and they also tend to be poorer on average. To me this seems an uncomfortably presumptuous hypothesis, especially for explaining the broader patterns of inequality in workplace safety documented in other settings. A more plausible possibility seems to be that it reflects subtle mechanisms of the marketplace, not dissimilar to the sorting in the housing market described earlier—a phenomenon that economists refer to as *occupational sorting*.[13]

Because safety might be an important component of total compensation, we might expect workers and employers to bargain for some combination of wages and safety that suits the objectives of each. For instance, if it is relatively cheap to add air-conditioning to a series of 100-square-foot offices, but more expensive to do so for a 40,000-square-foot warehouse, employers may offer air-conditioning as a standard amenity to office workers but not for warehouse workers. Instead, they may offer warehouse workers slightly higher pay (*compensating differentials*) relative to what similarly qualified individuals might otherwise receive in the next best alternative (for instance, working as a bartender or a truck driver). In other words, all else being equal, workers in more exposed and thus more dangerous or unpleasant conditions would be paid a wage premium.

The jury is still out on whether workers are indeed paid compensating differentials for climate risks. Importantly, available evidence

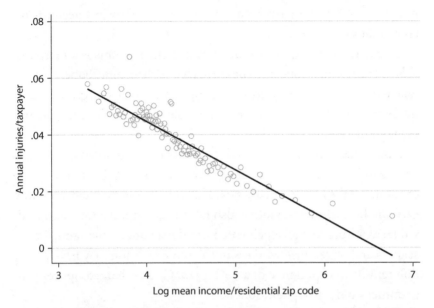

FIGURE 9.2. The relationship between the rate of workplace injuries and log residential income by zip code in California from 2000 to 2018. Workplace injury data comes from approximately 11 million workers' compensation claims. Average residential income data comes from the IRS. *Source*: R. Jisung Park, Nora Pankratz, and A. Patrick Behrer, "Temperature, Workplace Safety, and Labor Market Inequality" (IZA Discussion Paper no. 14560, July 2021), https://docs.iza.org/dp14560.pdf.

suggests that all else is often not equal, and that other economic forces may push this in the opposite direction. As mentioned above, while economic theory predicts a pay premium for dangerous work, it also predicts that workers with different resource endowments may make different trade-offs, including those regarding safety risks on the job, leading to a positive relationship between material living standards and occupational safety.

Consider figure 9.2. It plots the average workplace injury rate by average income level by zip code of residence, using data from California once again. It suggests that, on average, poorer people work in more dangerous occupations.[14] Even if some such workers are compensated with hazard pay on the margin, it seems that lower-income workers are more likely to be willing to fill dangerous roles, in part because they may have fewer safer alternatives, and in part because they just need the income that much more direly.

To use the language introduced in this chapter, this is a matter of *vulnerability*, which appears to be correlated with income. Poorer people tend to work in baseline more dangerous jobs, which for a given level of heat exposure makes them more vulnerable to climate hazards, including workplace injury.

It bears mentioning that overall exposure may vary in ways that contribute to this pattern. We saw above how, even within the same city, the dynamics of the housing market tend to funnel poorer people into more extreme local climates. Notice also how, even within a location, individual exposure to a given climate hazard may not be the same. Landscapers and warehouse workers in the same location may have a very different lived experience of a 90°F (32.2°C) day than computer programmers and nurse practitioners.

These and other highly local factors may lead poorer individuals to face both greater exposure and greater vulnerability to climate hazards. If these factors hold true generally, we might conclude that climate damages will be unequally distributed in ways that exacerbate observed patterns of income inequality.

Zooming out, one important implication of these examples is that climate change may hit harder not only for poorer countries, but for poorer people in all countries—even in richer, developed nations. Growing evidence suggests that, in addition to large North–South inequality in potential climate damages, there may be significant variation in climate damages within countries, even across individuals within specific neighborhoods and companies and families.

How Climate Change and Poverty May Interact

For most upper-middle-class Americans, the decision of whether or not to turn on the air conditioner on a very hot day is typically a no-brainer. Granted, there are those who worry about the potential planetary consequences of running the air-conditioning, or the opprobrium of environmentally conscious neighbors and colleagues. Nevertheless, it is typically not a decision that involves serious deliberation about the consequences for financial solvency. Most have enough of a cushion or access to credit that allows them to easily smooth out unexpected expenses, including a higher-than-usual electricity bill.

For the more than 30 million households in the United States who live on less than $35,000 a year, this may be a much bigger deal. This isn't necessarily because the heat itself is insurmountable per se. More that, for some, it bites in the context of very little financial cushion to work with and may force harsher trade-offs.

Economists Jacqueline Doremus, Irene Jacqz, and Sarah Johnston tested this idea using household survey data from across the United States. They found that lower-income households were much less responsive in their energy use to extreme temperature, even in places where everyone has air-conditioning. Everyone uses more energy on very hot and very cold days. But the poor, who have less slack in their budgets, respond more sparingly.

Importantly, they find that poorer households respond to extreme temperatures by cutting back on other spending as well, including on necessities like food. This is not the case for richer households, whose food expenditure does not vary in response to hotter or colder temperatures.[15] This is important given that as of 2020, even in a rich country like the United States, over 13 million households (about 10 percent) were food insecure or had difficulty providing food for all family members because of a lack of money.[16]

This is consistent with the idea that, for those whose budgets are strapped to begin with, even less extreme weather events—a slightly hotter day here, an unusually cold night there—can trigger qualitatively different responses relative to those with more slack to work with.

Alan Barreca, Paul Stainier, and I wanted to see how the financial distress of something as seemingly mundane as a few hot days might extend for poorer American households. We were motivated by the observation that many lower-income households report in interviews that they engage in precarious "juggling" to be able to pay utility bills on time, including taking out expensive payday loans, or reducing purchases of other essential items such as medicine in order to avoid delinquent payment.[17] For many such households, the threat of having one's electricity disconnected may loom ominously given the disruption it can cause.

We were able to partner with a major utility to access individual account-level data for over 300,000 low-income households in California.[18] Pairing each account's monthly payment and disconnection status with daily weather data near the home, we were able to study how temperature affected not only the household's electricity bill, but also their likelihood of disconnection. We found that even a few very hot days could, in some cases, tip the scales toward bill delinquency that ultimately led to disconnection.

In our data, each additional day with a maximum temperature of 95°F (35°C) causes electricity expenses to increase by 1.6 percent in the current billing period. So a week above 95°F increases monthly electricity bills by around 8 to 10 percent.[19]

These added expenses appear to be enough to tip the scales for some households. Hotter temperatures lead to a statistically significant increase in disconnection for these poorer households. An additional day with highs around 95°F increased the relative risk of disconnection by 1.2 percent. This means that, when faced with a week of temperatures above 95°F, poor households are 5 to 10 percent more likely to be disconnected from their electricity service, possibly due to bill delinquency.

It is worth noting that, despite the contemporaneous impact of heat on energy expenditures, the impact on actual disconnections does not appear to be immediate, likely due to policies that prohibit utilities from disconnecting customers without prior notice. In our data, the effects on disconnection showed up with a lag time of roughly two months after the increased expenditures.

For households already in poverty and stretched financially thin, even small additions to their budget can trigger a cascade of harsh trade-offs. Our findings are consistent with sociological interviews in which poorer individuals experience significant psychological stress over the fear of possible disconnection. They are perhaps more important considering the fact that a nontrivial fraction (over 5 percent) of US households report taking out payday loans—often at interest rates of over 300 percent annually—to pay for necessities like food, rent, and energy.[20]

Thankfully, many disconnections in the United States are short. But such findings are also consistent with studies that find higher energy costs to lead to steeper health costs of extreme weather, due perhaps in part to precisely these forms of desperate trade-offs.[21]

A broader point is that, for the poor, even small climate shocks may lead to starker trade-offs. Notice that, in both cases, "poor" was in the context of household incomes below $35,000, which implies a per capita income of roughly $13,500 (assuming average US households have 2.6 people). This is still more than most people in the world. It is slightly less than the average Brazilian or Chinese; almost twice that of the average Indian; nearly five times that of the average Ugandan (all using PPP-adjusted figures).

Why Interpersonal Climate Inequality Matters

How do these more local climate inequalities affect the big picture? There are at least three reasons why inequality in climate vulnerability may be important to understand, not just at the international level but interpersonally as well.

First, averages in climate impacts can mask vast differences in extremes, especially in the context of significant economic inequality, in ways that affect our understanding of the aggregate damages to society. A world in which climate change mainly leads to increased mortality and nonmarket suffering for poorer people could, at least according to traditional measures of economic health like GDP or property value, appear as if it has sustained relatively little economic losses, if these losses are offset partially or fully by gains among populations in cooler climates (or muted statistically by negligible impacts on highly adapted richer individuals). Here, it is worth noting that even the established means of monetizing nonmarket damages such as health typically count the lives of poor people less.[22] This may have been one of the reasons why earlier cost-benefit analyses of climate change were less supportive of aggressive emissions reductions, an issue we will discuss in further detail in part IV.

Similarly, if climate change results in a significant spatial reallocation of income and economic activity within a country, the aggregate change for the nation may not be a great indicator of the social cost. Recent estimates of climate damages in the United States suggest that some of the poorest counties may see declines in total income that are roughly ten times larger than in the richest counties, and that even relatively small reductions in aggregate output may mask a significant transfer to the North and East from the South and West.[23] At a global level, the evidence increasingly indicates that global averages in terms of economic losses due to hotter temperature may mask significant heterogeneity in realized human harm, particularly in terms of the vast disparities in lives lost between northern and southern latitudes.

This in turn can alter our understanding of how serious the climate problem is and for whom, which may, depending on our personal and societal values, affect our understanding of the urgency of climate mitigation. We will save for chapter 11 the specifics of how quantifying differential damages translates into prescriptions for policy. For now, suffice to say that both the quantitative magnitude of the implied economic costs of climate change (often summarized as the *social cost of carbon*) and the qualitative criteria relevant for climate action (which may or may not include deontological ethical principles) may be affected by our understanding of the extent of inequality in climate damages. In other words, interpersonal inequality in vulnerability may affect how a given amount of warming translates into realized human suffering as well as the moral justifications for reducing greenhouse gas emissions.

Second, understanding the contours of interpersonal climate inequality may be important as a practical matter of political economy. This may be especially true in developed and middle-income economies that will need to engage in the bulk of emissions cuts over the following few decades. Of particular concern is the possibility that lower-income individuals may, feeling the twin pinch of higher energy costs and diminished economic prospects, further withdraw support for climate mitigation policies—even though some may also face accelerating if largely hidden climate damages. Most of the early action on mitigation will likely take place in richer economies—in places like the United States, the EU, South Korea, Australia, and Canada. What do we know about how climate change already affects lower-income populations in these countries? Assuming that it is possible to identify pain points faced by lower-income households, could it be helpful for politicians to communicate with specificity how climate change is contributing to their constituents' suffering, as well as how policies may help alleviate it?

Finally, given the amount of warming that is expected even with aggressive mitigation, we may wish to design proactive adaptation policies—particularly those that target the least advantaged. Here is where the why and how of interpersonal climate inequality becomes vital. In order for policies to actually help the people who need help most, and to do so in a cost-effective way, it will likely be important to have an understanding of not only the international contours of climate inequality—which countries are more affected—but also the specific reasons why some individuals are hurt more than others. Assessing whether and what kind of government intervention is desirable requires a clear understanding of *why* the poor are hit harder, and what, realistically, can be done to help them adapt cost-effectively.

New data and research is quickly recasting our understanding of the granular details of how climate change affects the rich and poor differently. This is a fast-evolving literature, one in which there remains many unresolved questions. This chapter has tried to make the case that the contours of climate inequality closer to home may be important to recognize.

10

The Hidden Determinants of Climate Vulnerability

Adaptation needs in the developing world are set to skyrocket to as much as $340 billion a year by 2030. Yet adaptation support today stands at less than one-tenth of that amount. The most vulnerable people and communities are paying the price. This is unacceptable.[1]

—UN SECRETARY-GENERAL ANTÓNIO GUTERRES

ONE HUNDRED billion dollars per year. This is the amount of funding that rich countries pledged in 2009 to help poorer nations adapt to climate change, largely in recognition of the fact that poorer countries are at once more vulnerable while having contributed less to the problem.

Some estimates suggest adaptation needs of the developing world may be much higher. According to a 2022 United Nations report, estimated annual adaptation needs may rise to $160–$340 billion by 2030 and $315–$565 billion by 2050.[2]

The question of whether these adaptation finance targets are adequate to the task is a complicated one. To start, it is difficult to clearly conceptualize how to measure climate adaptation costs because people have always adapted to climatic conditions, and, depending on factors like income and technology, the optimal level of adaptation to even a stable climate is not precisely defined. In principle, adaptation to climate *change* should be measured from this baseline of current adaptation, but delineating where current adaptation ends and adaptation to human-made climate change begins is a difficult problem, to say the least. Moreover, this difficulty is often compounded by the fact that, in many developing countries, separating the needs associated with climate adaptation from the needs associated with poverty is easier said than done.

Whether the magnitude of aspired and delivered adaptation finance flowing from Global North to South is enough is often contested. Less contested is the moral imperative behind some form of Global North–Global South resource transfer. Both the inequality in historical emissions and the inequality in economic outcomes that climate change is poised to possibly exacerbate make it difficult to disagree with the ethical persuasion of such arguments as that made by UN Secretary-General Guterres above.

But how often do we stop to consider what such funds should be used for? Should the money be transferred to the bank accounts of the world's poor? Setting aside the issue of whether those most in need have bank accounts to begin with, the magnitude of the adaptation challenges and the economic mobilization aimed at meeting them are likely such that questions about not only *who* is most vulnerable but *why* and how to choose among various adaptation policy options may become ever more important in a warming world.

Answers to many such pivotal questions have so far remained cloaked in uncertainty.

Income Matters

Central to the issue of global climate justice is the greater vulnerability of people in poor countries. One important determinant is the fact that poor countries tend to be much more reliant on agriculture. For poor countries in the 10th percentile of the global income distribution, employment in agriculture accounts for 65 percent of the workforce. For countries in the 90th percentile, this proportion is 3 percent.[3]

Crop yields can be highly sensitive to climate, and increased heat and precipitation variability are expected to have significant adverse impacts on agricultural production in many parts of the world. According to a recent metareview, 2°C of warming is expected to reduce yields of major staple crops—namely, maize, rice, soy, and wheat—by an average of 20 percent, with wheat and soy experiencing the most nonlinear effects. This is even though, on the margin, carbon fertilization from increased CO_2 pushes yields up, by around 11 percent for a doubling of CO_2 concentrations.[4]

Considering the likelihood that farmers in poor countries will typically have access to fewer adaptation options than has been observed in richer contexts, the effects of such yield reductions on livelihoods could be rather severe. This prompts an important question of what kinds of adaptation options are most cost-effective, and whether there are significant factors other than income that hinder their timely and widespread adoption. We will return to this question momentarily.

Another important piece of evidence indicating people in poor countries tend to be more vulnerable comes from studies of heat and human health. Returning to the study on heat and mortality by the Climate Impact Lab discussed in previous chapters, we can make a few additional observations. By estimating not only the causal effect of heat on mortality in the tens of thousands of regions they have data for, but also estimating how this effect varies by average income and climate, the researchers are able to estimate the role of income in potentially blunting the effects of future climate change.[5]

Their findings suggest that the relationship between hotter temperature and mortality depends heavily on income. Richer people tend to be substantially better insulated against the health risks associated with hotter temperatures, even when controlling for the average climate. In other words, a 90°F (32.2°C) day causes far more death in Bangkok, Thailand, than it does in Singapore due in part to the many adaptations that are afforded by higher incomes.

The Climate Impact Lab researchers estimate that, without any adaptation (so, simply extrapolating current dose-response relationships into a warmer future), climate change is likely to increase mortality rates by 221 per 100,000 people. As discussed previously, this is a massive increase that is over fifteen times the current US automobile fatality rate, and more than the reported US COVID-19 fatality rate at the height of the pandemic. But adaptation is projected to reduce this figure by almost two-thirds: from 221 per 100,000 to seventy-three per 100,000. The bulk of this—78 percent of the difference—comes from higher incomes.[6]

The second notable observation is that people in relatively hotter climates tend to be better adapted to heat, even after controlling for income. While income growth accounts for over 78 percent of the adaptations, the rest comes from better adaptations for a given level of personal income. This suggests that two persons living in equally affluent societies that have different long-run average climates (for instance, persons living in colder Poland relative to warmer Greece) have different sensitivities to heat-related health risks for reasons that are as yet unclear. In fact, even within the United States, the heat-mortality dose-response curve can vary enormously across cities like Seattle, Chicago, or Houston, which have different historical experiences with hotter temperature despite roughly similar average income levels.

A recent study by economist Ishan Nath suggests a similar pattern. He finds potential for highly effective adaptation in the context of manufacturing productivity as well, echoing some of the contrasts between the effect of temperature on manufacturing in the United States, Canada, and India noted previously. Using nationally representative firm-level data from seventeen different countries, including India, Brazil, the United Kingdom, and the United States, Nath finds that the effect of heat on worker productivity varies significantly with income. For the least adapted firms in poor countries with moderate climates, a day with a maximum temperature of 40°C (104°F) reduces annual output by 0.4 percent, which comes out to roughly the equivalent of erasing a full working day. In middle income countries, the impacts of a similarly hot day were roughly half as large, and in rich countries the effect was negligible on average, save for the least adapted firms in select areas.[7]

These and other similar findings suggest that income is a key determinant of vulnerability to climate change, or at least a highly informative proxy for it. It also suggests that a wide range of adaptations can be highly effective at blunting the adverse impacts of climate change, at least for weather events within the domain of observed experience. This is good news. The fact that adaptation can be so effective means that we have the opportunity to make whatever climate change we have locked ourselves into more manageable.

However, they also raise important questions. For example, *why* are poorer people more vulnerable? That the marginal effects of climate shocks are attenuated for higher-income groups does not tell us which of the many possible factors associated with being richer makes the biggest difference. In addition, we may ask why some adaptations occur with more experience living in a hotter climate, even without the benefit of higher incomes.

As mentioned at the outset, these questions are significant because as governments around the world become more serious about financing and implementing climate adaptation measures, it will make a big difference whether or not they can target such efforts toward the most effective solutions in an evidence-based manner. And, as we'll discuss in the final chapters, having a clear way of differentiating between adaptations that would be expected to arise naturally as the world warms (and as incomes rise, assuming future economic growth) versus those that require some form of government intervention to actualize may be increasingly important.

The Hidden Determinants of Climate Vulnerability

We know that, to a first approximation, climate change is going to hit hardest for subsistence farmers like Raj in the developing world. They are likely to be among the most exposed to various climate hazards, as well as the most vulnerable.

But a closer look at the specific constraints that lower-income farmers face in adapting to climate change reveal how easy it may be to overlook some of the most important points of potential leverage—the hidden institutional and economic fixes that may have an outsized impact on climate vulnerabilities for the world's poor.

A growing literature finds that constraints to adaptation in agricultural settings are not only about crops' inherent sensitivity to temperature and precipitation, but also about social and economic institutions that govern how farmers respond to climatic changes of various kinds.

For instance, a recent paper by economists Robin Burgess, Olivier Deschenes, Dave Donaldson, and Michael Greenstone suggests that access to banking may be an important fulcrum of adaptation and resilience in developing countries.[8]

Rural households in the developing world often lack access to even the most basic means of smoothing consumption over time. Without access to loans or other financial products, for instance, a few bad harvests might spell disaster. With them, households may have a much better chance of bouncing back and may be forced to make harsher trade-offs less often. An improved financial environment, such as one with better access to banking services, should in principle mitigate the impact of weather on well-being, including on health, to the extent that households are better able to sustain essential services and strain less during leaner times induced by poorer harvests.

Exploiting the rollout of India's rural bank branch expansion program, which ran from 1977 to 1990, the researchers found that access to local banking services substantially mitigated the impact of hot weather on mortality. The average district saw the impact of hot days on mortality fall by roughly 75 percent due to expanded bank penetration, from a 1.2 percent increase in mortality per day above 90°F (32.2°C) to a 0.3 percent increase. In addition, the authors found that hot weather sharply depressed agricultural yields and the wages of agricultural laborers in rural areas, suggesting that an important mechanism through which heat affects health in poorer rural areas may be through its indirect effect on livelihoods, in addition to its direct effect on human physiology.[9]

Access to insurance appears to play a similarly important role in reducing vulnerability. One study by economists Dean Karlan, Robert Osei, Isaac Osei-Akoto, and Christopher Udry found that Ghanaian farmers who were given access to weather insurance were able to not only reduce the downside risk associated with unpredictable and highly weather-dependent harvests, but to "crowd in" other investments—in other words, initial investment begetting more investment—that boosted their productivity.[10] Another recent study found that guaranteeing access to credit in the event of a flood not only led farmers to be more likely to overcome negative shocks when they happened, but also led them to make riskier but more profitable investments in their farms ex ante, possibly due to the added security that such credit guarantees provided. This is despite the fact that the loan product is financially viable for the lender as well.[11]

A third related factor is access to markets, domestic or global. Many in the developed world may take integration with national and international markets for granted. It is difficult to overstate the hypermarket connectivity enjoyed by most rich world consumers in a globalized, e-commerce age, where one can have myriad items delivered to one's door with a keystroke. For many poor people, especially in rural areas in developing countries, such market access is by no means a given. Whether due to lack of good roads, lack of trade, or lack of internet connection, many in the developing world face additional challenges in coping with climate shocks.

Whether we are aware of it or not, being plugged in to trade networks offers a form of implicit insurance against adverse food price shocks in that bad weather and reduced agricultural yields in one part of the world can be partially made up for by production in other places. This is not always the case for farmers in the developing world. Particularly for many landlocked African countries, a combination of high inland transport costs, high administrative fees, and long customs delays make trade additionally costly, even without accounting for tariff barriers. Such challenges can be compounded by what economists refer to as *the food problem*, whereby farmers in poorer countries with harsher climates (and thus less productive agriculture) who might otherwise benefit from specializing less in domestic agriculture, end up perversely specializing *more* as the world warms, in part because food is a necessity. According to work by Ishan Nath, this perverse overspecialization in low-productivity agriculture among the poorest, least trade-liberalized countries may lead to climate damages that are significantly larger than would be the case with greater openness and less reliance on domestic food production, particularly for poorer countries.[12]

Even when countries are well-integrated into international markets, many poorer economies may be additionally vulnerable due to differences in terms of trade, which refers to the ratio between the prices of a country's exports and the prices of its imports. According to research by economist Frances Moore and her colleagues, whether one is a net food importer or exporter matters. Because demand for food tends to be relatively inelastic, when global agricultural yields decline, food prices rise. For countries like the United States or Canada who tend to be net food exporters, this price effect may partially offset the decline in yields, leading to less economic damage. For countries like Nigeria or Egypt, who are net food importers, the price effect exacerbates the damage, making policies and systems that can serve as shock absorbers for residents of such countries all the more valuable.[13]

Shock Absorbers, Shock Amplifiers

In many agrarian societies, more rain during the growing season is a gift, as it can boost agricultural productivity. But whether this gift will ultimately be of benefit may depend on surprising factors like one's age and gender, and how the social and economic structures of the society one lives in adjust.

In a fascinating paper, economists Manishah Shah and Bryce Sternberg found that in rural India positive rainfall shocks led to *lower* human capital investments for most school-aged children, while they boosted later educational achievement of toddlers and infants, namely those age 2 and below and those in gestation at the time of the windfall of rainfall.[14]

How do we make sense of these seemingly conflicting findings? A few important contextual features stand out. As late as the 2010s, most rural Indians engaged in agriculture as their principal economic activity. This often includes children who may, depending on the economic environment, drop out of school to help on the farm. Moreover, due to limited access to international food markets, local agricultural productivity can be an important determinant of diet and nutrition in such settings. When times are good, more bellies are full. When times are bad, more mouths go hungry.

Given that much agriculture in India is rain-fed, more rain during the growing season can be a driver of both local wages and nutrition. This means that when positive rainfall shocks lead to bumper harvests, labor demand is high due to a greater need for hands to pick the crop. One consequence appears to be that many youths, particularly those old enough to provide meaningful labor, drop out of school during such periods. Because education features many path dependencies, such decisions can have persistent consequences, as evidenced by lower test scores and attendance rates in the immediate aftermath of a positive rain shock, and increased likelihood of dropping out of school or falling behind years later. Shah and Sternberg find that, for 5- to 16-year-olds, a positive rainfall shock decreases math test scores by 2–7 percent of a standard deviation and increases the chances of a child dropping out of school by 4 percent. The effects are particularly detrimental for 11- to 13-year-olds because they are most likely to be transitioning from primary to secondary school.[15]

On the contrary, toddlers, infants, and fetuses in gestation appear to benefit from the windfall, perhaps due to better nutrition and health—either for the children directly, or indirectly through improved maternal diet. For these groups, the researchers documented more educational investment later in life, as measured by higher test scores and higher chances of school enrollment many years later.

Findings like these underscore the importance of the economic and institutional environment in determining how climatic shocks ultimately affect well-being. Such factors can either serve as shock absorbers, reducing the realized harm, or they can act as shock amplifiers, exacerbating the long-term consequences of a given climatic event, and do so differentially across social groups. One can imagine very different dynamics playing out in a setting where children face less economic pressure to work, or one in which compulsory schooling and child labor laws are enforced more vigorously. It is worth bearing in mind the breadth of empirical evidence suggesting that early-childhood investments in education and health often pay large dividends over the course of a lifetime, even when the immediate opportunity costs of such investments (e.g., foregone household income) are high, as is often the case in poor, developing countries.[16]

These findings are also consistent with the idea that a complex and as yet somewhat opaque array of factors may connect climate change to human capital and educational outcomes more broadly, which ultimately may prove pivotal in determining whether the effects of climate change on economic development discussed earlier take the form of more damaging long-term growth rate effects. Recall the impacts of heat on learning discussed in previous chapters. Work by Teevrat Garg, Maulik Jagnani, and Vis Taraz found that hotter temperatures affect school performance in India through their indirect effects on nutrition and health, most likely operating through reduced agricultural yields.[17] Although in the long run investments in human capital will be a critical factor in whether developing countries are able to catch up economically, we still know uncomfortably little about the specific climate vulnerabilities faced by children in the developing world.

Another important set of institutional determinants of climate vulnerability may be found in the world of work. As we have seen, climate change is likely to expose workers to more potential workplace hazards. This suggests that how labor markets are organized and regulated will matter a great deal. Here too evidence is slim, but available indicators offer important clues.

According to the International Labour Organization (ILO), 60 percent of the global workforce is in the informal sector, contributing roughly 35 percent of global GDP. Sometimes referred to as *invisible workers*, these workers often don't have legal jobs or official contracts but contribute heavily to economic activity. Often though not always, garment makers, agricultural workers, and construction workers in developing countries fall into this camp.

Informal workers are less likely to have access to basic protections offered by social safety net programs including unemployment insurance or health insurance. They may be more likely to experience income volatility arising from climate shocks—agricultural workers without work during bad harvests, construction workers working fewer hours due to increased heat, and so on. In many countries, informal workers are also less likely to have access to insurance against workplace accidents, and more likely to be affected by workplace harassment, both of which have been shown to increase during hotter weather.[18] The role of work informality extends to our understanding of how climate change affects gender inequality as well. Women are estimated to perform six times the amount of unpaid labor as men, and in many countries they receive significantly lower wages within the same occupation and industry.

In countries like the United Arab Emirates, where temperatures routinely climb above 40°C (104°F), many of the most exposed jobs are manned by migrant workers, including in construction. Research by economist Suresh Naidu, Yaw Nyarko, and Shing-Yi Wang found that, due to restrictions on such workers' ability to find alternative employment outside of the firms who sponsor their visas, migrant workers are less able to bargain for higher wages. It stands to reason that such institutional factors may affect workers' exposure to climate hazards on the job as well.[19]

Will countries adopt changes in the labor market institutions and norms that govern the baseline protections workers are provided? These and other institutional factors may play outsized roles in determining how damaging climate change ends up being in the decades to come, for individual economic mobility and national economic development alike.

Innovating for the Bottom Billion

When I was in graduate school, I was asked to take part in a briefing for Bill Gates. It was part of a meeting held in Seattle aimed at bringing Gates up to speed on the latest climate change research. Much to my surprise, I, along with a few other academics, would spend an entire afternoon with Mr. Gates himself. No fancy entourage. No cameras. Just Bill and a handful of us talking data.

Our session was framed largely in terms of how the emerging evidence influenced the Bill and Melinda Gates Foundation's understanding of the need for swift climate mitigation. But in retrospect, I wish we had focused more on the role that philanthropists like Bill Gates needed to play in facilitating climate adaptation.

The reason isn't simply because climate change looks to be a force that could reverse a significant fraction of the gains in livelihoods made among the world's bottom billion. It is because, in studying the issue of adaptation since then, it has become increasingly evident that there are gaps in the adaptation marketplace that only coordinated global efforts can fill. This is perhaps especially true in the case of incentivizing innovation for adaptation technologies and practices that might help the world's poorest individuals.

Technological change is not a magical process that happens spontaneously; it is a long and expensive process of tinkering and investment, influenced profoundly by the economic incentives surrounding would-be inventors. In the energy sector, we have already achieved remarkable redirection of technical change toward cleaner technologies like renewables through subsidies, societal norms around environmental sustainability, social pressure by activists, and changes in norms around the process of technological change. The same coordinated effort may be needed for pro-poor adaptation technologies, especially those that could benefit the world's bottom billion.

Global philanthropy may have a critical role to play in catalyzing the development and distribution of breakthrough adaptation technology. An analogy to a malaria vaccine might be apt. Because so many of the people who would benefit most from such a breakthrough have limited financial and political voice, and because they are diffuse across many different jurisdictions, each with limited incentive or wherewithal to make the necessary R&D push, the private market had been unable to provide an effective vaccine. Perhaps that is why the Bill and Melinda Gates Foundation donated over $250 million in 2005 toward malaria vaccine research.

Similarly, in the absence of a global government, there may be a gap to be filled by private philanthropy in developing next-generation low-cost, low-emissions adaptation technologies.

What might this technology look like? As an economist and not an engineer, I can only speculate. It could be a solar-powered heat pump for rural homes, or a battery-powered wearable cooling garment for outdoor workers, or something totally unexpected. Evaporative cooling or cool roofs for less humid hot areas might also be an option.

It might even be that, to a first approximation, the technological solution to facilitating adaptation in the world's poorest places has little to do with physical tech and more to do with financial innovations that help the poor gain better access to markets and capital.

The point is that philanthropists like Bill Gates may have a significant role to play in facilitating climate adaptation, including in incentivizing innovation for adaptation technologies and practices that can help the world's poorest individuals. The private market may be recruited as partners in such endeavors through public-private partnerships, but it is unlikely to fill this gap on its own. This is particularly important considering recent evidence that innovation in the climate adaptation space appears to be relatively limited. One analysis using global patent data found that the share of inventions pertaining to climate adaptation in 2015 was roughly the same as those in the year 1995.[20] The lack of progress in research and development for adaptation is in stark contrast to the growth of climate change mitigation technologies, which saw their portion of overall innovation more than double over the same period.

Climate Vulnerability and Climate Justice

If we cannot stop the seas from rising . . . if you allow for a two degree rise in temperature, you are actually agreeing to kill us.

—MOHAMAD NASHEED

Mohamad Nasheen, the former president of the Maldives, a small island nation in the Indian Ocean, once held a cabinet meeting underwater to demonstrate the potentially catastrophic future consequences of climate change for low-lying island nations like his own.

He and the leaders of other small island nations have made repeated pleas to the international community to limit warming to the more ambitious target of 1.5°C. The crux of their argument is that the eventual sea level rise associated with such warming would likely wipe many of these countries out of existence, and that most of the emissions contributing to such sea level rise have come from richer, developed countries. Their demands are part of a broader narrative that emphasizes the culpability of the Global North and the aggrievement of the Global South.

The international community, in addition to making 1.5°C an aspirational target and pledging to provide significant adaptation finance as mentioned earlier, has recently also agreed to entertain the concept of *loss and damage*, which could add to North–South financial flows under the aegis of climate justice in a significant way.

Small island nations provide a particularly poignant case of how multiple layers of inequality and injustice could overlap when it comes to climate change. Many readers might agree that there are elements of the situation that are deeply unfair.

I have suggested that whether or not one is inclined to be devoted to the mission of climate justice, an important starting point may be a dispassionate and fact-based understanding of *who* is likely to be hurt most by a changing climate, and *why*. While it may be true that climate change brings into harsh relief many historical injustices, particularly between the Global North and the Global South, an overemphasis on moral indignation at past wrongdoing risks obscuring the very action-relevant gaps in our collective understanding of the problem at hand, particularly with respect to reducing vulnerabilities.

As I have tried to highlight here, there are at least two dimensions of climate justice that risk being obscured amid the familiar narrative. The first is the highly local way in which climate damage may vary, and how this informs our understanding of the optimal targeting of support. It is true that, to a first-order approximation, the biggest winners and losers from climate change will likely be defined in terms of whether one resides in a rich or poor country. But increasingly, the data suggests that, at least over the medium-term, climate change may cause significant hardship for poorer *people*, even in relatively affluent countries like the United States, South Korea, France, Germany, and perhaps most importantly in middle income countries like China, India, and Brazil. Conversely, richer individuals outside of the developed world—whether they be the urban elites of Shanghai or New Delhi or

Accra—are likely to be shielded reasonably well from many of climate change's initial blows.

The second is that we still know uncomfortably little about *why* some groups are hurt more than others and how the adverse impacts of climate change on one group of people ripples out across the global economy through its many complex interlinkages. This is important in part because such information may be critical in helping citizens, governments, and philanthropists decide what kinds of adaptation interventions might be most effective at reducing harm. Of course, as ammunition for climate rallies, the *whether* of unequal impact may be sufficient information. But once we take seriously the challenge of evidence-based climate adaptation policy, we can begin to appreciate how many gaps in our collective knowledge base there are.

To summarize, climate retribution may be an important component of what it means for us to live with dignity in a just world. But the familiar trope of righteous indignation risks missing important dimensions of the problem.

Returning to the story that started this section, it may be easy to focus on the fact that Niran and Raj are simply poorer and will experience more intense climate change than James. It's also clearly the case that Raj, being a subsistence farmer, will probably be hit even harder by a given increase in climate hazards. These facts, while important, are only part of the story.

There is much we do not yet understand about how climate change affects the poor. In general, having higher incomes helps protect against climate change. But why exactly that is the case remains something of a mystery. An improved understanding of which adaptation solutions should be prioritized and to whom may be just as important as the amount of financial resources available to poorer countries. In some cases, simply boosting incomes by way of cash transfers may do the trick. In others, more systematic overhauls to the social and economic infrastructure of a community may be first order. These may include improving the availability of local credit and banking services or other less obviously climate-related interventions like reforming educational institutions or labor market policies.

As developing countries receive international climate financing for adaptation and begin to expand their efforts at facilitating adaptation and resilience, understanding how these can be best allocated to high-impact regions and policies becomes an increasingly urgent priority. Of course, important caveats to this technocratic perspective deserve mention. History, including very recent history, cautions against the hubris of interventionalist agendas, particularly by a Western, technocratic elite.[21] As the specter of colonialism still looms large in many developing and formerly developing countries, it would be wise to approach the communities that we seek to help with genuine humility and a sensitivity to the procedural dimensions of justice as well. Still, the evidence I've presented suggests that the more we can learn about what works and what doesn't when it comes to climate adaptation, the better chance we might have in helping to protect the world's most vulnerable from the warming we have in store.

PART IV

WE HAVE EXPLORED how climate change may be damaging to society because of its subtler, slower burning consequences. Among other things, this perspective reveals the importance of paying more careful attention to the social and economic determinants of climate vulnerability and risk.

How might it inform practical decisions moving forward? One can imagine a wide range of choices that could benefit from incorporating the slow burns and their determinants into the calculus: individuals deciding whether and where to purchase a home; businesses and investors deciding where to locate facilities and allocate capital or whether to invest in new production techniques; lawmakers deciding how best to regulate markets, raise revenue, or engage in public works projects; or individual citizens deciding whom to vote for in upcoming elections. While it is impossible to provide a playbook for all or even most such choices, my hope is that the chapters that follow can provide a useful framework for thinking about how climate change factors into them.

I will focus on the implications for two types of policy interventions: climate mitigation and climate adaptation. The former pertains to how we can reduce greenhouse gas emissions, while the latter focuses on making our economies and societies more resilient to the warming that is now inevitable.

Of course, discussions of policy solutions are rarely straightforward because they often require moving from descriptive statements to normative ones. This means that the data and research findings must be interpreted through a set of decision criteria. These criteria may encompass how to prioritize current versus future cash flows, how one values the well-being of less advantaged (or even nonhuman) members of society, or whether we should judge the appropriateness of an action based on its outcomes or some inherent moral principle.

Ultimately, the evaluation of climate policy options will depend in part on a weighing not only of various costs and benefits, but also of personal and collective values, including regarding the appropriate ethical principles to employ. In what follows, I hope to provide an economic framework for thinking about how empirical data might inform policy decisions moving forward, emphasizing the difference between private (internalized) and social (externalized) costs in a manner that is consistent with a wide range of value frameworks and ethical positions.

While I may at times allude to different conceptions of what constitutes a "just" society, the idea is not to champion a particular ethical position but rather to note how the data described may serve as an input to decision-making tools commonly used by policymakers, including cost-benefit analysis, while acknowledging a pluralism of possible values and ethical frameworks. It must therefore be noted at the outset that, by adopting an economic framework for evaluating policies, I will implicitly be adopting a set of ethical and philosophical principles—including welfarism and consequentialism—that judges policies based primarily on realized outcomes, including potentially the distribution of realized economic outcomes, and that reasonable minds can disagree with. But it is not my intention to champion a particular stance, for instance, on how economic resources should be distributed across members of society. Nor must this approach exclude the application of other criteria that, for instance, place more explicit emphasis on duties, moral absolutes, intentions, or individual liberty.

11

Never Too Late

IN 2019, the British newspaper *The Guardian* decided to do away with the term "climate change" from their reporting altogether, replacing it with terms like "climate emergency," "climate crisis," or "climate breakdown." Explaining the shift, its editor-in-chief pointed out that the term climate change "sounds rather passive and gentle when what scientists are talking about is a catastrophe for humanity."[1]

Or, as a joint statement from a group of outlets (including *The Guardian*) put it: "Why 'emergency'? Because words matter. To preserve a livable planet, humanity must take action immediately. Failure to slash the amount of carbon dioxide in the atmosphere . . . could render a significant portion of the Earth uninhabitable."[2]

Words do matter. And as global discussion of climate change has become more and more focused on the concept of climate catastrophe, there has been a concomitant—though of course not necessarily causal—rise in what some have begun to call *climate anxiety*.

A 2020 survey of 2,000 young people aged 8 to 16 in the United Kingdom found that more than 73 percent of respondents were worried about the state of the planet. A total of 41 percent did not trust adults to tackle the challenges posed by climate change, and 19 percent had experienced a bad dream about climate change.[3] Another survey of 16- to 25-year-olds in ten countries including Australia, Brazil, France, India, Nigeria, the Philippines, the United States, and the United Kingdom, found that over half of all respondents reported feeling sad, anxious, angry, powerless, and guilty, so much so that over 45 percent reported such feelings about climate change negatively affected their "daily life and functioning."[4]

Of course, it is difficult to assess whether the rise in feelings of climate-related fear and anxiety are a result of the way we discuss the issue, given the multitude of other correlated factors, including rising social media use or the actual increase in extreme weather events, both of which may affect mental health directly.

And yet, it's hard to blame anyone—young or old—for getting depressed by climate change given the proliferation of dystopian accounts of it. One particularly gripping portrayal is provided by "The Uninhabitable Earth" by David Wallace-Wells, which debuted as a *New York* magazine article in 2017 and later became a *New York Times* bestseller. Describing desolate future landscapes of unlivable cities, "the end of food," and human civilization in shambles, Wallace-Wells admonished his readers that "no matter how well-informed you are, you are surely not alarmed enough."[5] The fact that much of the conversation has focused on the symbolic target of 1.5°C, despite the fact that there is no bright red line that separates a world with 1.4°C of warming and one with 1.6°C, has certainly not helped with public feelings of fatalism and doom.[6]

The possibility of irreversibly melting ice sheets that set in motion long-run sea level rise that could inundate the world's major cities is indeed terrifying, even if most estimates are that the long run in such instances refers to at least a few hundred years from now, if not a millennium.[7] Even the terminology we use—"runaway warming," "unlivable cities," and "biblical floods"—can be anxiety provoking in itself.

On the one hand, there are very good reasons why we might want to view the climate change problem through the lens of insurance against planetary catastrophe. The earth's climate systems may feature important tipping points that, if crossed, could set in motion unknown amounts of additional warming over the longer term. To name a few potential tipping points that you may have come across: there is the possible melting of the Greenland Ice Sheet, which could lead to significant additional sea level rise; the possible shutting down of the thermohaline circulation system in the North Atlantic (imagined in the movie *The Day after Tomorrow*); dieback of the Amazon rainforest, which could result in a self-perpetuating cycle of reduced rainfall, dieback, and carbon release; and the discharge of methane from arctic permafrost, among others.

There are also good reasons to suspect, as many of the most vocal champions of the climate catastrophe narrative seem to have done, that traditional economic frameworks and tools—in particular, cost-benefit analysis—may be ill-equipped for assessing potentially catastrophic, intergenerational problems like climate change. Knightian uncertainty, sometimes also referred to as *deep uncertainty*, is one of them.[8] The issue isn't simply that there is a probability distribution around our projections of the future, it's also that the basis on which we can construct such probabilities is highly limited to begin with, making it difficult to interpret the results of probabilistic cost-benefit analysis, sort of as if we were rolling a strange, lopsided die of unknown origin. Furthermore, there are legitimate concerns about the ability of economics to incorporate nonmarket aspects of human and ecological health and well-being into cost-benefit analyses of climate change, not to mention sufficiently capture the distribution of costs and benefits that are so important in thinking about equity and fairness. The fact that many of the important ecological impacts may be irreversible also presents challenges for standard cost-benefit analysis.

An intellectually responsible assessment of how to respond to climate change is therefore not possible without grappling with the deep uncertainty and irreversibility associated with anthropogenic warming, and the possibility, however small, that our inaction may lead to something that is truly catastrophic for civilization as we know it. Such possibilities likely place bounds on the usefulness of traditional economic tools of policy evaluation, like cost-benefit analysis, particularly without ample contextualization and supplementation.

On the other hand, the "climate change as looming cataclysm" framing, in isolation, also misses important pieces of the picture. In this book, I have tried to present an alternative framing of the climate change problem, one that focuses on the often hidden, slow burn. One that fixates less on the 200-year global calamity hypothetical and more on the diversity of more seemingly minor yet cumulatively important challenges posed by an incrementally warmer world, today and over the coming decades. Throughout, the intention has been to offer this view as an important complement to the catastrophe insurance narrative, not necessarily as a substitute.

And yet, it seems increasingly important that more voters and decision-makers add the slow burn perspective to their heuristic toolkit. There are many reasons for this, as we have explored throughout the book. Another reason arises paradoxically from the progress that we've managed to achieve on climate mitigation.

Noncatastrophic Climate Damages

In the late 2000s, middle-of-the-road projections by the Intergovernmental Panel on Climate Change (IPCC) put the likely range of warming somewhere between 3°C and 5°C by 2100.[9]

Best estimates now put us on track for a business-as-usual warming trajectory between 2°C and 3°C, though with uncomfortably uncertain tails in the full distribution of potential outcomes. If nations follow through on the pledges they have made in international negotiations to date, this appears likely to be closer to 2.1–2.4°C.[10] Once again, it is worth stressing the myriad uncertain factors that complicate even these updated end-of-century projections, whether it be about temperature or the realized impacts of temperature rise. Sticking with temperature projections for a moment, it may well turn out that melting permafrost and boreal forest dieback releases far more planet-warming methane than scientists anticipate today, or that reduced albedo toward the poles accelerates warming more quickly than models currently predict.

Nevertheless, it seems reasonable to suppose that thanks in part to pledged and actual emissions cuts achieved in the past few decades, the likelihood of truly disastrous warming may have declined nontrivially. Cataloging the full suite of reasons for this revision of forecasts is beyond the scope of our present analysis. Some are related to surprisingly rapid advances in mitigation technology, and others to improvements in the scientific community's ability to model the earth's climate and economic systems, including how much coal we will end up using under business as usual.

This shift in the likely trajectory of climate change is arguably something to take to heart. At least in terms of projected warming, the prognosis has certainly improved. Some might even characterize it as partial success.

Far from making understanding the economic consequences of climate change less important, this partial progress arguably makes a nuanced understanding of climate damages more valuable. It places increased importance on the task of differentiating the multiple shades of undesirable between climate normal and climate apocalypse, as well as the challenge of sifting through the messy socioeconomic details of climate adaptation.

This is where the "climate catastrophe" framing carries with it an additional risk. It is the risk that, once it becomes clear that the path of warming we are currently on is not so obviously catastrophic, climate change ceases to be a problem worthy of scarce mental bandwidth and political capital.

Here is an illustration of this point. Five years after "The Uninhabitable Earth," Wallace-Wells penned a follow-up piece in which he made a dramatic about-face. In "Beyond Catastrophe: A New Climate Reality Is Coming into View," the author argued that "truly apocalyptic warming now looks considerably less likely than it did."

Wallace-Wells goes on to write that "the past few years . . . have made me more mindful of the inescapable challenge of uncertainty when it comes to projecting the future and the necessity of nevertheless operating within it," noting that his earlier alarmism, while well-intended, may have been overly pessimistic.[11]

Personally, I find Wallace-Wells's attempt at reining in previous temptations of apocalyptic thinking to be commendable. In fact, much of the parenthetical nuance of his earlier message seems to have been lost in the fray. For instance, in "Uninhabitable Earth" he reminds readers early and often that "annihilation is only the very thin tail of warming's long bell curve." Nevertheless, this episode underscores the risk of motivating climate action through fears of climate apocalypse alone.

It also points to a danger associated with not having a disciplined way of quantifying climate damages as a spectrum occupying many shades of gray, rather than being a binary black and white—climate catastrophe or climate normal.

The approach and the evidence discussed in this book provide at least one important set of tools to help make more transparent and scientifically disciplined, if still highly uncertain and imperfect, assessments of how bad the climate problem is and how aggressively we might want to act on it.

As we will discuss in this chapter, it turns out that both approaches—that of catastrophic disaster insurance and measuring the costs of the slow burn—now appear to point toward the same general conclusion in terms of emissions reduction targets that the global community may want to set in the present. From the perspective of *sustaining* costly emissions cuts in the future, during what will likely be a decades-long transition, it is perhaps equally important to better understand the noncatastrophic consequences of climate change, partly because, as societies become more serious about decarbonization and the costs of this effort begin to mount, questions will continue to arise as to why such costly effort continues to be necessary.

Just as the camera crews' focus on the wildfire's flames keeps us from appreciating the dangers of the lingering smoke, a focus on the bright red line of whether there is a planetary catastrophe or not risks inuring us to the very real suffering that noncatastrophic warming will engender. And just as our fascination with the flames may be keeping us from taking concrete steps to minimize our exposure to the smoke—recall the casual resumption of campus activities while the Skirball fire still smoldered—it seems possible that a fever pitch around climate emergency may have discouraged inquiry into the specific ways in which societies may want to prepare for the warming we have in store.

Changing Tides

In the choppy waters separating Great Britain and Denmark lies the Hornsea 2 wind farm. A regiment of more than 150 colossal wind turbines jutting out of the sea, it boasts a generation capacity of over 1.3

gigawatts, rendering Hornsea 2 the world's most extensive commercial wind farm. Plans are to raise that capacity to six gigawatts by the time construction wraps up for all six Hornsea projects, matching the capacity of six nuclear power plants and providing enough electricity to supply 4.5 million homes.

A short sail north is the North Sea oil field, an underwater archipelago of ancient fossil fuel deposits that has yielded tens of billions of barrels of oil since the late 1800s. As offshore wind farms and their associated industries have gained prominence, oil production in the region has waned. In 1999, British oil production in the area exceeded 125 million tons (6 million barrels) per year. In 2022, that number dwindled to 35 million tons, less than one-third of what it was two decades ago. In coastal cities within reach, whether in Denmark or the United Kingdom, labor demand increasingly skews toward cleaner industries, with employment in oil and gas extraction continuing to fall as positions in the wind sector rise.[12]

As with any industry, numerous factors are likely at play in the fluctuation of fortunes. These may include global market trends in oil demand or changes in production costs—such as the expense of tapping increasingly hard-to-reach fossil fuel reserves or the efficiency of wind turbine blades in converting kinetic energy into electricity. They may also include the availability of surplus workers who have lost their jobs due to globalization, or the influence of geopolitical events such as the Russia–Ukraine war.

However, a force that has weighed increasingly heavily on energy markets are the stick and carrot of climate change policy, manifested in both anticipatory voluntary measures and binding regulation. The Hornsea 2 undertaking is too costly to be a mere public relations stunt or the outcome of an environmental, social, and corporate governance (ESG) initiative. Its driving force has always been the profit motive, underpinned in part by the assurance of high and rising permit prices in the EU and UK emissions trading schemes, the prospect of tax credits and subsidies for renewable energy, and growing consumer demand for cleaner energy to power their products and commutes.[13]

As we will see, one figure encapsulates the combined effect of many of these forces. This number, while perhaps not scrutinized as closely as economic indicators like the Federal Funds Rate or the price of a barrel of oil, is likely to exert a significant influence on investment activity in the years ahead. The number in question is the social cost of carbon.[14]

Not unlike the way a star's mass determines its gravitational pull on neighboring planets, one could say that, in an evidence-based policy environment, the magnitude of the social cost of carbon represents the velocity and force of the coming shift toward clean energy.

Energy firms such as BP, which have a stake in both the polluting and clean energy sectors, have been utilizing an internal social cost of carbon to inform their investment strategies for many years. According to a 2021 McKinsey report, over 40 percent of energy companies surveyed said they were using an internal social cost of carbon to guide investment decisions, a proportion that is likely to rise in the years to come, even for companies not obviously tied to the energy sector.[15]

Markets, Government, and Climate Mitgation

Before we talk about the social cost of carbon, it is worth revisiting a fundamental concept in economics. What follows is a short primer on the economics of climate change policy, specifically, the framework many economists use in thinking about government intervention in the context of greenhouse gas mitigation. Readers who are well-versed in the basics of how public economics addresses the case for government intervention in the economy may wish to skip ahead to the next section of this chapter.

What is the proper role of government in addressing climate change? One would be forgiven for finding this question a bit perplexing. There seems to be a lot of blame to go around, including toward ineffectual policymakers as well as big bad business. "For 50 years, governments have failed to act on climate change. No more excuses" reads a 2022 op-ed in The Guardian. Meanwhile, a growing number of corporations have made heavily promoted climate pledges, ostensibly obviating the need for government intervention in a newly socially responsible corporate landscape. "Are you ready for the ESG revolution?" reads a 2021 report by PwC Global, a major consultancy, on the heels of major corporate sustainability announcements, including Apple's commitment to become carbon neutral by 2030, "20 years sooner than IPCC targets."[16]

Despite these seemingly mixed messages, whether government action is necessary in reigning in carbon emissions is not really a cause for debate. At least among economists, experts across the political spectrum agree on the principle of government intervention in the context of environmental pollution, including staunch libertarians.

Milton Friedman, a Nobel Prize–winning economist who has been referred to as the "most revered champion of free-market economics since Adam Smith," argued that there is one and only one social responsibility of business: "to use its resources to increase its profits while engaging in open and free competition without deception or fraud."[17] At least in the public lexicon, Friedman is often associated with a strong stance against government encroachment on the free market. When asked in 1979 what he would do to put the economy back on track, he responded, "Cut down government spending—to hold down the rate of growth of government spending in dollars and to cut it in terms of purchasing power," a position echoed in many of his writings.[18]

Friedman notes elsewhere, however, including in his influential book *Capitalism and Freedom*, that there are clear cases where the government must step in "to do something that the market cannot do for itself." These include what he terms "neighborhood effects," that is, "when actions of individuals have effects on other individuals for which it is not feasible to charge or recompense them." He goes on to list environmental pollution, "an obvious example being the pollution of a stream," as a clear case of such neighborhood effects warranting government intervention.[19]

This is largely because of the first fundamental welfare theorem, which is as fundamental to microeconomic theory as Newton's laws of motion are to physics. It states the conditions under which competitive markets ensure efficient allocation of resources: namely, perfect information, no transaction costs, no market power, and no externalities. If any of these conditions are violated, then the free market cannot necessarily be relied on to deliver a socially efficient outcome. (Efficiency here refers to societal resources being utilized to their fullest potential, including environmental resources such as clean air or climate stability, in this case from the standpoint of humanity writ large.) In the case of externalized costs, positive or negative, the free market will do a poor job of creating outcomes that are beneficial for society, as in the case of the negative externality that is carbon pollution. Something similar applies to monopolies, where firms can wield their market power to raise prices and reduce production, making outsized profits at the expense of the greater good.

To correct these externalized costs, some form of policy intervention is often necessary. What form of policy intervention this should take is a separate question, the answer to which the first fundamental welfare theorem provides little guidance. Even if the existence of market failures makes it possible for government intervention to improve societal well-being, unless policies are designed and implemented carefully, this improvement is by no means guaranteed.

Indeed, perhaps one of Friedman's most powerful observations may have been in recognizing the significant informational requirements for government intervention—even in such cases as "neighborhood effects" or monopolies—to meaningfully improve on the existing situation. Among Friedman's many insights is that much the same way a botched surgery can make a bad health situation worse, well-intentioned policy initiatives can be implemented in ways that replace existing problems with new ones. This is arguably a reason why there is considerably more disagreement regarding both the optimal *level* of government intervention in climate change, and the ideal *implementation* mechanisms, whether it be a carbon tax, a cap-and-trade program, technology subsidies, or something else. This speaks to the difference between the economics of the justification for some form of policy intervention and the merits of alternative policy instruments, which is a feature of the debate that can sometimes be conflated as well.

To summarize, there appears to be overwhelming consensus within economics that the market cannot solve climate change on its own, even if the market will need to play an important role in arriving at the solution.[20] To a first approximation, a most useful framework for thinking about climate policy is in terms of externalized costs, which can be thought of as an application of Friedman's "neighborhood effects" and the first fundamental welfare theorem. According to this framework, the case for government intervention in terms of correcting the climate externality is clear, even if, on the margin, there are firms and consumers willing to voluntarily clean up their act without the threat of government intervention.

Or, as economist Geoffrey Heal puts it in his illuminating book *When Principles Pay*: "The goal of government regulation can be viewed as attempts at bringing private costs more into line with greater social costs."[21] Even though some firms may find it in their interests to commit to carbon neutrality or other climate-friendly objectives voluntarily, for the average firm, the magnitude of the misalignment of incentives (i.e., the externalized costs) is likely such that, without government intervention, society will experience too much climate change.

Once it is determined that there are significant externalized costs, we can seek empirical guidance as to how much correction is needed and which kinds of interventions—some more expensive than others—may be justified. One of the ways we can do so is by quantifying the magnitude of externalized costs and including them in standard cost-benefit analyses of government policies and investment decisions. In the case of climate change, this guidance primarily takes the form of estimating the social cost of carbon.

The Social Cost of Carbon Revisited

The social cost of carbon is a concept that tries to help us engage in such cost-benefit comparisons systematically, by representing the severity of climate impacts as a dollar figure. Technically, it is the discounted present value of all avoided future damages associated with reducing greenhouse gas emissions by one metric ton today.[22] Practically, it is perhaps one of the most influential numbers in the context of climate policy because it gives us a way of conceiving the societal costs of non-catastrophic climate change and because it will likely dictate the direction and flow of many billions, if not trillions, of dollars in investment.

The social cost of carbon can help decision-makers engage in an apples-to-apples comparison when deciding how much to invest in clean energy and other carbon-reducing technologies. One way to think about it is as a tool that helps *unstack* the deck so that the interests of future generations and the world's poor are represented in economic decisions. Because there are several different greenhouse gases, including but not limited to carbon dioxide, the term *social cost of greenhouse gases* is also used. As economist Michael Greenstone puts it: "The social cost of carbon was needed because otherwise the benefits will always be measured in tons of carbon, and the costs in dollars, and the dollars will always win."[23] It is a concept that is as controversial as it is powerful. Within the field of economics, there is an active debate about what should and should not go into the calculation of the social cost of carbon, and how such a figure should be interpreted and used in policy-making. More broadly, there continues to be debate over whether important nonmarket climate damages can be realistically quantified, or whether the act of putting dollar values on something like a human life or biodiversity writ large is morally appropriate.

That being so, in government and in the private sector, the social cost of carbon already plays an important practical role. In government, it plays a growing role in determining whether certain rules can be instituted, and how stringently regulations are enforced. In the United States alone, there have been over eighty regulations that have been justified based on the social cost of carbon. These include rules governing energy efficiency of consumer appliances, fuel-economy standards for vehicles, and the phasing out of hydrofluorocarbons, which are used in refrigerators and air-conditioning.[24]

The social cost of carbon plays an increasingly important role in guiding private investment decisions—and not just in fossil fuel–related companies. One estimate suggests that nearly 30 percent of the world's top 100 financial services companies, and over 15 percent of travel, logistics, and infrastructure companies have already adopted an internal social cost of carbon to guide decision-making, with more planning to do so.[25] Novartis, a Swiss-based global healthcare company, uses a carbon price of $100 per ton of CO_2 in determining which projects to pursue. Microsoft has pledged to make its operations carbon neutral through a carbon fee modeled after the social cost of carbon; within the company, each business group is responsible for paying the fee depending on how much energy it uses.[26]

All of this means that the scientific trends affecting the calculation of the social cost of carbon may be of significance far beyond the confines of academia or the technocratic circles of government ministries.

Earlier models that calculated the social cost of carbon made significant simplifications due to data limitations. For example, many integrated assessment models treated the whole world as one economy, using a single global damage function or damage functions for only sixteen different regions. A damage function is, in essence, a function that relates a given amount of warming (e.g., 1.5°C, 2°C) to monetized damages. As we've seen, averages can mask important variation, particularly when looking at data on such aggregate measures as national or global GDP. Because many earlier models extrapolated damage functions estimated using highly aggregated data from rich countries to infer damages for the entire world, their outputs suggested a relatively low social cost of carbon, and thus a less urgent case for mitigating emissions.

Earlier models also made strong assumptions about the sectors that were likely to be affected. For instance, Nordhaus's Dynamic Integrated Climate-Economy (DICE) model assumed in 1991 that trade, manufacturing, finance, and other sectors—collectively responsible for 87 percent of total economic output at the time—were insulated from climate change impacts because they occurred indoors or were otherwise "negligibly affected."[27] The Climate Framework for Uncertainty, Negotiation and Distribution (FUND) model, which is another major integrated assessment model (IAM) often cited in rulemaking, assumed that global agricultural output would actually increase in the context of moderate amounts of warming, due in part to using agronomic model-based estimates of potential yield (as opposed to realized damages using empirical data) and extrapolating carbon fertilization effects. We know based on the studies discussed earlier that sectors like trade and manufacturing, while perhaps not affected as directly or intensely by climate shocks as agriculture or construction, are nevertheless significantly affected by adverse weather already, though uncertainties around future adaptation remain. We also know that the sensitivity of agriculture to heat and drought are likely to be much higher than these earlier models assumed.

These limitations arose in part from lack of good data, particularly in the construction of climate damage functions, resulting in improvised assumptions about the presence or absence of relationships. As we've seen, our understanding of the relationship between climate and economic activity has evolved significantly in recent years. Research in this area has grown rapidly, especially since 2010, combining big data and causal inference to inform the shape and constituent parts of the climate damage function in much greater detail. In parallel, a rich literature has clarified many aspects of discounting for large-scale intergenerational problems like climate change, which has helped clarify another controversial dimension of the quantification of climate damages.

All of this means that our ability to zero in on the true social cost of carbon has improved significantly since then by incorporating more data-driven and credibly causal inputs to the damage function, accounting for the fact that many aspects of human well-being respond more sharply to higher temperatures than previously thought, using real-world data to simulate future adaptation.

The general trend has been for the new data to revise our estimates of the social cost of carbon upward. In 1991, economist William Nordhaus estimated the social cost of carbon associated with a ton of emissions in 2015 to be roughly $4.50 per ton, based on the influential DICE model. He and his colleagues later revised this estimate to $31 per ton in 2015, using an updated version of the DICE model. Under the Obama administration, the US government estimated the social cost of carbon to be $40 per ton in 2015, on the basis of output from Nordhaus's models as well as two other influential IAMs (the aforementioned FUND model and the Policy Analysis of the Greenhouse Effect [PAGE] model).[28]

The most recent update by the US Interagency Working Group places the social cost of carbon closer to $190 per ton.[29] This upward evolution can be attributed to several factors. Here, I will focus on three. These include the updating of damage functions that now incorporate more empirical inputs and pay closer attention to adaptation, improvements in long-run socioeconomic forecasts, and refinements in our understanding of the conceptually appropriate discount rate.

Previous models appear to have significantly underestimated the increase in mortality due to higher temperatures. For instance, the Climate Impact Lab estimates an increase of 73 deaths per 100,000 by the end of the century, compared to the 0.33 deaths per 100,000 estimated by the FUND model. This represents an increase in the slope of this dimension of the climate damage function of over 200-fold.

These upward revisions in the slope of the damage function are in some ways even more surprising given that they also incorporate adaptation explicitly, accounting for the fact that most places will be significantly better able to deal with climate change in the future, which would put downward pressure on overall social cost of carbon estimates. As we noted before, the authors of the Climate Impact Lab study account for the fact that most places will be significantly richer and better adapted to hotter temperatures in the future, thus reducing the total impact of heat on mortality. And yet, despite this empirical inclusion of adaptation, the mortality consequences of climate change are projected to dwarf earlier estimates by two orders of magnitude or more.

Additionally, the earlier models provided limited scope for nonmarket damages, including potential effects on mortality, morbidity, labor, crime, learning, and other less tangible aspects of economic activity. While more recent estimates have yet to incorporate many of these nonmarket damage mechanisms, by including effects on channels such as mortality, labor supply, and energy use, they begin to provide a more comprehensive account of potential climate damages.

Over time, the income and population projections that form the basis of the models' estimates have also been refined. These projections are important given the long timescales involved in cost-benefit analyses of climate change. They directly influence expected future emissions trajectories and affect aggregate climate damages because more people and more expensive property at risk means the same proportional damages will result in more aggregate losses. The latest models now employ econometrically generated probability distributions of future income and population trajectories, which have brought much-needed empirical nuance to these crucial modeling inputs.

Importantly, our understanding of discounting and how standard practices of discounting future benefits and costs in policy analysis should be amended in the context of highly uncertain, intergenerational problems like climate change have evolved significantly in the past few decades as well.[30]

You may be familiar with the idea that discounting played a central role in earlier economic debates regarding the social cost of carbon. In perhaps one of the more publicized academic debates on the issue, economist Nicolas Stern was heavily criticized by other economists, including William Nordhaus, for his choice of discount rate in writing the influential *Stern Review on the Economics of Climate Change*, published in 2006.[31]

The *Stern Review* was notable in that it argued for far more aggressive emissions cuts than was conventional wisdom among economists at the time. Critics like Nordhaus argued that this conclusion was driven largely by the choice of discount rate, which in the case of the *Stern Review* was significantly lower than many previous models, including Nordhaus's.[32]

Nordhaus put it this way: "The Review's unambiguous conclusions about the need for extreme immediate action will not survive the substitution of assumptions that are more consistent with today's marketplace real interest rates and savings rates. Hence, the central questions about global-warming policy—how much, how fast, and how costly—remain open."[33] Stern and others pushed back, arguing that the ethical premises upon which one could apply such high market-based discount rates to global and intergenerational policy problems were slim.

Much convergence has occurred in the ensuing years. While far from a settled debate, there is now much greater agreement in the literature over what the relevant discount rate should be than in the immediate aftermath of the *Stern Review*. A recent survey of experts on the determinants of long-run discount rates found the median response to be 2 percent, and that 92 percent of experts were comfortable with a social discount rate in the range of 1 percent to 3 percent.[34]

This is a clear departure from earlier models estimating the social cost of carbon that used much higher discount rates, including 6 percent. The reasons for this shift are complex and manifold but include growing consensus among experts that a morally defensible pure rate of time preference for social discount rates—that is, the rate at which society's impatience discounts the value of benefits and costs arising to future generations—is likely quite close to zero; the observed decline in average risk-free market interest rates over the past thirty years; the recognition that government investment in one area may crowd-out other public investment, in which case a shadow cost of capital is the right metric to use, not the market rate of return, as in earlier models; as well as growing consensus over the idea that, in the context of uncertainty, the right discount rate is a declining schedule rather than a single fixed rate.[35]

It is worth noting that for the most part, changes to discounting in analyses of climate change are mostly technical in nature, rather than ethical. While the recommended discount rates have trended lower, it is not necessarily because researchers have collectively decided to prioritize the welfare of future generations. Rather, this trend can be attributed primarily to theoretical and empirical aspects of the social discount rate, including our understanding of risk and future economic growth projections. This is important in light of our intention, expressed earlier, of providing a value-transparent (if not fully value-neutral) exposition of how the data informs decision-making related to climate change policy.

These refinements suggest a significant revision in our understanding of the economic costs of climate change, as well as the stringency of the policy response that is merited. A social cost of carbon of $190 per ton has vastly different implications for the scale and pace of clean energy investments than one of $30 or $50 per ton.

It is of course important to caveat that some aspects of the social cost of carbon remain highly debated. There are still many factors that are missing, many of which could result in significant revisions. Current estimates, while now better incorporating nonmarket damages like the effects of temperature on human health, fail to consider other nonmarket mechanisms, such as potential effects on human capital and learning, morbidity, crime, and negative affect as well as potential losses to ecosystems and biodiversity.

The fact that, in the updates noted above, nearly half of the entire social cost of carbon is driven by temperature-related mortality, a nonmarket climate damage, is highly notable. This suggests that further expanding the scope of nonmarket damages that are included in the social cost of carbon may be important in ensuring that it reflects as close to the full set of damages as possible. Once again, it is worth putting the relative magnitude of this less salient slow burn into perspective. In recent estimates of the social cost of carbon, sea level rise accounts for less than 3 percent of the total damages, and changes in energy costs are estimated to be negligible, in part because of offsetting changes in heating costs that are expected to decrease, and cooling costs that are expected to increase.[36]

Furthermore, the latest estimates fall short of fully incorporating how unequal impacts may manifest differently in terms of welfare losses (that is, changes in realized well-being) across the world. Even though the latest models do a better job of characterizing the inequality in marginal climate damages—how heat affects health differently for people in Germany versus Ghana, for instance—they do not go so far as to ascribe different value weights to benefits that accrue for richer or poorer populations. This is because they do not engage in *equity weighting*, which adjusts for the fact that a dollar's worth of damages to someone living on $2 a day has a different realized impact on well-being than on someone earning $200 a day. While equity weighting is not yet standard practice in government cost-benefit analyses in most countries (Germany is a notable exception), it may be important to consider given the magnitude of inequality in climate impacts, as well as evidence that human beings generally tend to value additional consumption at a diminishing rate, which is consistent with the idea of equity weighting.[37]

As mentioned previously, many have criticized IAMs for doing a poor job of representing the kind of uncertainty inherent to a problem like climate change. This critique still holds. Because of the long time horizons and degree of uncertainty—including regarding the set of potential outcomes and their probabilities, also referred to as deep uncertainty—the expected utility framework that undergirds IAMs rests on shakier ground than for many other policy problems for which cost-benefit analysis is typically used. This means that their outputs may be misleadingly precise, and that, when using such measures as the social cost of carbon to guide policy, these contextual factors must be kept squarely in view.

A related aspect of uncertainty has to do with how we incorporate individuals' and society's preferences regarding risk into the calculation. If you were told that, in the event that you get sick under your current insurance plan, the medical bills associated with treatment would be between $200 and $1,000, with an average scenario of $500, you might be willing to forgo paying more for additional coverage. If instead you were told that the cost of treatment may vary more widely—say between $50 and $20,000—though with the same average scenario of $500, this might increase your willingness to pay for additional insurance to reduce the risk of being saddled with an unaffordable medical bill. If so, the implication is that your preferences exhibit some degree of risk aversion.

Research increasingly indicates that, if individuals are risk averse, incorporating uncertainty can substantially increase the implied social cost of carbon. Whatever probabilistic damages we estimate to be the median scenario, adding uncertainty likely means that our willingness to avoid the entire distribution of outcomes increases, thereby increasing the implied social cost of carbon and making the case for climate mitigation stronger. Including such possibilities remains high on the list of to-dos for researchers in this space, especially given the risk of tipping points in the climate system.

And yet, even here, progress is being made, leveraging many of the tools and findings discussed previously. Some researchers, including Simon Dietz and his coauthors, have attempted to quantify the impacts of known tipping points in the climate system, including thawing permafrost, ice sheet disintegration, and Amazon dieback. They estimate that incorporating such tipping points could increase the central estimate of the social cost of carbon by around 25 percent, with a 10 percent higher probability of a social cost of carbon that is double the estimate without tipping points. This suggests that the existence of tipping points increases the overall risk associated with failing to act on climate change. It is also consistent with the idea that crossing tipping point thresholds does not automatically connote civilizational annihilation.

While these and other thus far omitted factors would likely push in the direction of an even larger social cost of carbon, some factors, such as the opening up of new shipping routes in the arctic or the reduction of energy costs associated with heating during winter months, could have a positive impact on the climate damage function and push in the opposite direction. And, as we have seen, changes in average warming and its overall costs can mask significant variation across geographies and persons, including at highly local levels.

To summarize, one way to interpret the upward revision in the social cost of carbon—which arises in part from more data-driven characterization of the climate damage function and a more principled (and data-driven) approach to discounting—is that climate change is a very big deal even if it does not turn out to be the end of the world, and even if what one mostly cares about is the economic cost to humans.

Of course, it may turn out that, to a first order, the most important benefits of the greenhouse gas reducing actions we take are to reduce the likelihood of planetary collapse. It may conversely be that, depending on one's values, the most important reason to reduce emissions is because of the equity and social justice dimensions of climate change, or because of climate change's implications for biodiversity and the welfare of wildlife.

The point I'd like to stress here is that, even if one takes a strictly human-centric view and focuses on the more readily quantifiable slow burns for the hypothetical representative global citizen, the data now suggests that we may be setting in motion significant economic harm to our collective well-being now and for many generations to come, justifying aggressive emissions cuts on economic considerations alone.

Trillions per Year, and Counting

How much harm are we causing by continuing to emit greenhouse gas emissions? One way to answer this question, albeit as a crude approximation, is to take the new estimates of the social cost of carbon at face value and approximate the additional damage we are causing each year in dollar terms.

Globally, we emit around thirty-five gigatons of CO_2 equivalent each year, adding to the stock of greenhouse gases in the atmosphere and increasing global temperatures. Since 1850, we have emitted approximately 2,500 gigatons of CO_2 equivalent, and the resulting stock of greenhouse gases is responsible for the warming we are experiencing. Given that the amount of warming depends on the cumulative concentration of greenhouse gases, rather than the annual flow, it is not unreasonable to assume that a year's worth of emissions does not significantly alter the trajectory of warming, which is worth remembering for the ballpark calculation that follows.

Suppose we were on track to reach 3–5°C of warming by 2100 without any greenhouse gas mitigation. Then the additional damages from emitting more greenhouse gases today would be the future damages associated with this warming trajectory. The $190 per ton value of the social cost of carbon would imply that the present discounted value of the damages from emitting one ton of CO_2 in 2022 is approximately $190 (using a 2 percent social discount rate). If we assume for the sake of simplicity that marginal damages are roughly linear, this would imply that our greenhouse gas emissions may be setting in motion damages valued at roughly $6.65 trillion each year ($190 multiplied by 35 billion). Of course, it is quite possible that this overstates the actual effect of a year's emissions—if, for instance, damages are nonlinear in overall greenhouse gas concentrations, or the business-as-usual trajectory is no longer so steep in expected warming. Nevertheless, a social cost of $190 at current emissions rates most likely implies additional damages from current emissions somewhere in the range of trillions of dollars per year.

To put this in perspective, global GDP in 2021 was $96 trillion, and the US GDP was $23 trillion. This implies that the present value of climate damages associated with carbon pollution could be as much as 6 percent of global GDP and 30 percent of US GDP.

How do these damages compare to the costs of reducing emissions? An important conceptual device in understanding mitigation costs is the marginal abatement cost curve, which represents various mitigation options and their potential scale (e.g., how much avoided deforestation is possible), arrayed in descending order of their per ton cost of abatement. Global marginal abatement cost estimates vary, and each carry with them important assumptions. It is worth noting also that some costs are very low or even negative. For instance, in some contexts, replacing coal power with shale gas is economical even without a price on carbon or other government support for low-carbon alternatives. Similarly, blending small amounts of corn ethanol into gasoline (up to 10 percent) as an octane booster can, at least in the United States, pay for itself. This would suggest that many emissions cuts might occur even without government action, irrespective of the social cost of carbon.

However, the general pattern is that while some amount of emissions reductions can be economically effective even without a price on carbon, in order to achieve mitigation that gives us a realistic chance of preventing 3°C or more of warming requires unlocking many investments, like electrifying automotive transport or fully replacing coal and natural gas with renewable energy even in places where the sun doesn't shine as brightly or where the wind doesn't blow as strongly, which at least at present represent a significant additional cost for the same level of service delivery.

It is also important to note that such marginal abatement cost curves typically represent *static*—as opposed to *dynamic*—costs in that they do not account for the potential for current investments to reduce the costs of similar technologies over time. From the perspective of an individual firm or investor today, the static costs are perhaps closer to the decision-relevant ones. But from the perspective of the global social planner, which is arguably the perspective one should take in evaluating the question of how much humanity would like to prioritize climate mitigation, the right costs should reflect dynamic costs, since we would be taking into account not only the costs and benefits for the present day but for future generations as well.

Setting this caveat about static versus dynamic costs aside for a moment, let us assess current *static* cost estimates and compare them to the damages mentioned above. Goldman Sachs publishes estimates for marginal abatement costs through their Carbonomics platform. Let us take these estimates as illustrative for our thought experiment. Their figures suggest that achieving 50 percent global emissions cuts relative to today would cost approximately $0.7 trillion per year. Ratcheting up emissions cuts to 75 percent would cost something on the order of $3.1 trillion per year.[38]

One interpretation of the numbers then is that, given current estimates of the social cost of carbon ($6.5 trillion in benefits, i.e., avoided damages per year) and mitigation costs associated with reducing emissions ($3.1 trillion in costs per year to achieve 75 percent reductions), mitigating global emissions by at least 75 percent makes investment sense from a purely economic perspective. Of course, this is not the same thing as making sense purely from a private *business* perspective. As mentioned before, the trouble is that without regulation that corrects for the free-rider problem and negative externality aspect of emitting greenhouse gas emissions, no individual firm or investor has the incentive to make these choices, which is what necessitates some form of government policies, ideally benchmarked to an evidence-based estimate of the true social cost of carbon.

There may of course be factors that increase abatement costs, including global military conflict, unexpectedly high implementation costs of low-carbon technologies, or unexpected bottlenecks in critical minerals used in the production of energy storage devices such as electric vehicle (EV) batteries, and so on. Such factors might push in the other direction, arguing against throttling up emissions reductions too aggressively.

On the other hand, if the dynamic costs associated with many cleaner technologies are lower than their static costs, the economic case for aggressive mitigation may be stronger than the above thought experiment implies. For instance, economies of scale and learning by doing may mean that there are positive externalities associated with investing in or consuming a particular form of nascent technology today because these expenditures may lead to engineering and managerial improvements in the future, which help to bring down the technology's costs over time.

An Optimism Born of Incrementalism

Regardless of the specific mitigation path chosen, addressing climate change will likely be more of a marathon than a sprint. The policy actions taken, and the investments already made, might be thought of as a large down payment on a giant multidecade mortgage. We, our children, and our grandchildren will likely need to make recurring payments to sustain the energy transition over time.

Why such a long time frame? One may wonder whether we cannot simply rip decarbonization off like a band-aid and get it over with. To start, there are significant transition costs. We cannot push pause on the world economy while its myriad carbon-intensive components are upgraded to cleaner ones. Food must be grown and shipped and prepared and served. Commutes must be taken, dwellings heated and cooled, hospitals and schools and factories and server farms kept in operation.

There is also the physics of the climate system to contend with. As mentioned already, climate change is a stock problem, not a flow problem. The amount of warming we experience depends not on the *flow* of emissions into the atmosphere in any given year, but on the total accumulated *stock*. As with the depth of water in a bathtub, even if the faucet is eventually closed completely (zero flow), we are still stuck with water from the past (accumulated stock) unless there is a way to speed up the drainage, which would be akin to negative emissions.

This means that, if we wish to reverse warming that has already occurred in a reasonable time frame, not only would we need to reduce the flow of emissions to zero—global carbon neutrality—we would also need to increase the rate of removal of greenhouse gases from the atmosphere. Some of this is already happening in the background thanks to uptake by the oceans and the growth of plants. But to make a meaningful dent on the stock of cumulative emissions, we may need to deploy various carbon capture and storage technologies.

Many such technologies are now technically feasible, and some believe it is only a matter of time before they become economically competitive. But the bottom line is that there is a long way to go between current emissions and carbon neutrality, let alone *negative* emissions.

And yet, there is a growing case for optimism. As already mentioned, the world has already corrected course nontrivially, especially relative to a few decades ago. In the European Union, total emissions have fallen from a peak of nearly four gigatons CO_2 to less than 2.8 gigatons in 2022, a decline of nearly 30 percent. Emissions per capita have fallen from eleven tons per person in 1990 to less than seven tons per person in 2022, a reduction of over 35 percent.[39] In 2021, the Tesla Model 3 was the second most popular new car in the United Kingdom. Not the second most popular EV, the second most popular new car, period.[40] And in Norway during the same year, roughly eight in ten new passenger vehicles sold were all-electric.[41]

Over a similar period, the United States achieved a less impressive but still significant 8 percent reduction in CO_2 emissions (1990–2020), and with the passage of the Inflation Reduction Act has put itself on track to nearly halve emissions relative to 2005 by the year 2030. Even among developing countries, there have been significant gains despite the continued upward march of emissions. In India, more new solar and wind power has been added in the past fifteen years than in Australia, Britain, and the Netherlands combined.[42]

Much of this progress reflects changes in market signals, both in terms of falling costs of cleaner technologies and rising penalties for dirtier ones. They also most likely represent voluntary anticipation of stronger binding regulations to come. Between 2010 and 2022, the cost of solar power fell by over 85 percent. The cost of wind power fell by 60–70 percent. The cost of lithium-ion batteries has fallen to around a tenth of what it used to be.

While attributing the causes of such declines is yet challenging, evidence increasingly points to important, mutually reinforcing roles played by policy and private entrepreneurship. For instance, a recent study by economist Todd Gerarden estimated that, without solar subsidies to consumers led initially by Germany's solar feed-in tariffs, the amount of solar power generation adopted in the period 2010–2015 would have been 51 to 78 percent less than what was the case. In effect, subsidies for early adoption of cleaner technologies helped to level the playing field in favor of solar over dirtier alternatives and may have generated spillovers of the kind mentioned above.

These are but a small sampling of the developments that are moving the world toward a less carbon-intensive economy, and thus a less rapid pace of warming. Studies like Gerarden's hint at the possibility that, for many clean technologies that are still ramping up, like EVs, the dynamic costs may be lower than the static ones.

The number of countries that have implemented binding carbon legislation—whether in the form of a carbon tax, a cap-and-trade system, or some combination of mandates and technology standards—has grown by the year. Whether driven by anticipation of regulation, changes in consumer sentiment, or investor pressure, more and more companies have also begun to take aggressive action. These and other developments suggest that the clean energy train has left the station and is quickly gathering steam. As noted previously however, much challenging work lies ahead, especially if we are to limit warming to something in the vicinity of 1.5–2°C.

When it comes to the question of whether we can solve climate change, the evidence presented in this book provides what is at once a more sobering and more hopeful picture. It suggests that even moderate amounts of warming may be far more harmful than we previously thought. At the same time, it suggests that even if some tipping points may have been crossed, it is never too late to make a difference. Every increment of warming prevented by the world today can mean billions if not trillions of dollars in damage avoided tomorrow.

Perhaps there was always a risk of drawing a bright red line in the sand to motivate action; it is the risk of needing to find a way to continue the hard work of emissions cuts once the line has been crossed, to rally toward a messy muddling through. If it is indeed true that our efforts to date have significantly reduced the risk of global catastrophe, then in order to sustain the effort needed to bend the emissions curve further, we will need a more quantitatively nuanced understanding of the paths before us—a way to more credibly delineate the social costs of a +3°C world and a +2.5°C world, versus a +2°C world or a +1.5°C one.

The quantification of climate damages, however imprecise and limited, gives us a clearer representation of not only the dire risks of not reducing greenhouse gas emissions more aggressively, but also the very tangible benefits of every increment of mitigation achieved, whether that is from a future likely involving 2.5°C of warming or 2.4°C, or 2.3°C, and so on. This, I would argue, is one of the best cases that can be made for informed and sustained optimism in the realm of climate mitigation policy.

12

Beyond Silver Bullets

TO THE EAST OF LONDON, a panorama of steel and concrete stretches across the Thames, forming a shield against surging tides. Guided by precision engineering, the Thames River Barrier can be raised to stand sentinel against the encroaching tides or gently lowered to allow the free flow of river traffic. For decades it has protected London from the threat of floods, which have menaced the city since Roman times.

Across the Atlantic, echoes of Hurricane Sandy's impact on New York City continue to reverberate. Now more than a decade after the 2012 storm, memories of destruction and loss still haunt the consciousness of a city scarred by the hurricane's unexpected intensity. As part of the ongoing response, the US Army Corps of Engineers has become immersed in an exploration of a similarly mammoth flood defense system—one with an estimated price tag of $119 billion.[1]

The proposed sea wall would consist of man-made islands with retractable gates, stretching from the Rockaways in Queens to a strip of land in New Jersey south of Staten Island. This six-mile-long wall would serve as a barrier to prevent damage from future flooding of the region as climate change causes sea levels to rise and storm surges to intensify.

The New York City sea wall has generated intense debate among residents and policymakers as they grapple with the larger question of how societies should adapt to climate change.

These kinds of expensive public works projects are enticing as climate adaptation options. Creating high-tech physical shields to guard against some of climate change's blows provides a certain psychological appeal. But could such ambitious proposals distract us from some of the real investments we need to be making? Captivating as they may be, could they divert our attention from the more pressing investments in adaptation that demand our immediate focus—from the messier, more mundane fixes to the way we handle mortgage insurance, the way we provide weather forecasts, or the protections we provide to workers?

These may seem like technocratic non sequiturs, rather removed from the mainstream climate conversation. Even the discussion of urban sea defenses, if you are not a resident of a major coastal city, may seem like someone else's problem. Except that it probably is not. Purely from a financial perspective, it is most likely everyone's problem. For instance, of the $119 billion, the proposal being considered would have New York and New Jersey paying 35 percent and Congress footing the other 65 percent of the bill. If we assume rather conservatively that all 20.1 million residents of the New York–Newark–Jersey City metropolitan statistical area are served by this project, this would amount to a per-person subsidy by the federal government of $3,848, likely paid for by the tax dollars of most Americans.

Will We Adapt?

For many years, discussing the question of whether humanity will adapt to a changing climate was almost taboo. Even when it was entertained, the issue often felt like an afterthought, something brought up mainly as a proxy for disciplinary turf wars or ideological disagreements. This may have been because, at its core, the debate over adaptation was still tethered to the larger question of whether the climate problem should be taken seriously at all.[2] Indeed, some proponents of adaptation used the prospect of technological fixes such as air-conditioning to mitigate the effects of heat or drought-resistant crops to address changing rainfall patterns to cast doubt on the idea that climate change posed any real threat.[3]

As we come to accept that a significant amount of warming is inevitable, and that the costs of this warming on both human society and the natural world may be large, the relevant question becomes less about *whether* we will adapt, and more about *how quickly, in what ways,* and *at what cost.*

As climate-related disruptions become more intense and public understanding of physical climate risks grows, debate over the role of government in facilitating climate adaptation will likely intensify. We've already explored how the effects of hotter temperatures on society, though often not very visible, are already pervasive, subtly affecting everything from physical and mental health to labor productivity, student achievement, violence, migration, and crime. Even if one is not personally exposed to high levels of climate risk, it is likely that policy decisions about adaptation will affect one's own pocketbook indirectly because most government efforts will be funded via tax revenues.

"Fiscal spillovers," like the tax burden associated with the New York City sea wall proposal, are especially important to consider given the vital roles that governments already play in spreading risk across time and space through their implicit or explicit status as insurer of last resort. For instance, in the United States, upward of 95 percent of flood insurance policies are insured through the Federal National Flood Insurance Program. In countries such as Spain, New Zealand, and France, national governments mandate comprehensive natural disaster insurance and provide explicit public reinsurance backstops to private insurers. In both cases, taxpayers across the country effectively subsidize homeowners and businesses in the most climate hazard-prone areas of the country in direct and indirect ways.[4]

At an international level, we've discussed how developed countries have agreed to global adaptation finance of $100 billion per year and possibly more to help developing countries cope with a changing climate. This raises important questions of how this money should be spent.

This chapter is about the coming adaptation rush—the multitude of decisions big and small that will be required in adapting to a warming world, and which, if made wisely and in an evidence-based way, have the potential to greatly reduce the harm and suffering that climate change will ultimately cause.

In this final chapter, I will argue that, increasingly, we have the tools to make adaptation policy in a principled and evidenced-based way. Yet again this requires checking our impulse to focus on the more visible ailments and to favor the more familiar remedies. Or, as one observer put it, to resist the temptation of the "shiny object, the silver-bullet fix luring us away from where need to go."[5]

Corporations and Climate Change: Adapting to Physical Climate Risk

Awareness is rapidly changing, and I believe we are on the edge of a fundamental reshaping of finance. . . . The evidence on climate risk is compelling investors to reassess core assumptions about modern finance.

—LARRY FINK, CEO OF BLACKROCK

While New Yorkers debate the sea wall and other proposed public adaptations, tides are shifting within the corporate sphere. In response to rising awareness of climate risk, companies are increasingly seeking better information that may help them uncover hidden vulnerabilities to the bottom line, whether in their own operations or further up the supply chain.

Climate risk can, from a financial perspective, be broadly divided into two categories. There is *transition risk*, which is the risk to businesses and investors associated with society's mitigation responses to the threat of global warming. (Related concepts include *stranded assets*, which broadly refer to the risk of broad classes of financial assets becoming stranded as the economy pivots away from fossil-fuel based production and consumption, as well as subcomponents of *regulatory risk* having to do with regulations aimed at facilitating the transition to a low-carbon economy.) Then there is *physical climate risk*, which is the risk to a company's assets, operations, and revenue streams associated with ongoing and projected changes in the physical climate. This is sometimes further subdivided into acute and chronic physical risks.

Historically, investors and corporate executives appear to have been more focused on transition risk—that is, when they were focused on climate risk at all. This is perhaps especially true among investors with significant financial positions in fossil-fuel-intensive industries and companies whose business models have relied heavily on cheap energy (e.g., steel, cement) or fossil-fuel based transportation systems (e.g., internal combustion engine automotives, aviation).

Increasingly, decision-makers across a wide range of industries are also tuning in to the latter—to physical climate risk. According to a 2020 survey of CEOs and CFOs, over three-quarters said their firms are not fully prepared for the adverse financial impact of a changing climate. Floods, droughts, and wildfires were listed as the three exposures that "could most negatively affect their financials" and that "concern their companies the most."[6]

Managers are asking such questions as: How exposed are the company's assets to physical climate risk? What are the major vulnerabilities in the supply chain? How can operations be made more robust to the increasing frequency and intensity of extreme weather events? Particularly for larger, multinational corporations, answering such questions skillfully may mean millions of dollars of losses averted and may even lead to new business opportunities that have been overlooked.

Exploring the incentives facing firms in their pursuit of climate resiliency may be instructive regarding the question of when government intervention in climate adaptation is and is not warranted. The evidence presented in this book suggests that the gradual and prolonged impacts of climate change may carry greater risks for companies than commonly assumed, potentially leading to significant financial consequences. For instance, it suggests that the chronic, often hidden impacts of hotter temperature may constitute a financially material form of physical climate risk, and that some of the most important threats may not be from the headline-grabbing disasters but from the complex propagation of a series of smaller impacts accumulating in not so obvious ways. At the same time, economic theory suggests that it isn't always obvious whether adapting to such risks is something the government should be responsible for.

Hotter Temperature and Supply-Chain Spillovers

Consider the hypothetical but not implausible situation facing a multinational athletic apparel company we will call Springbok. With North American headquarters in New York and a large European presence centered around Frankfurt, it is renowned for designing high-quality athletic shoes, sporting equipment, and apparel, which it distributes to thousands of stores worldwide.

Recognizing the long-term effects of climate change on both business and society, its C-suite was among the first to implement ambitious initiatives to reduce emissions and foster innovation of climate-friendly products and processes. Lauded as a pioneer and industry leader in the fight against climate change, the company has instituted programs that allow consumers to buy offsets for the emissions arising from the production and distribution of the shoes they purchase, and purchase agreements with power companies that allow it to operate almost exclusively on renewable electricity.

How vulnerable might Springbok's core lines of business be to a warmer world? At first glance, not very. Most of the company's employees are based in Western Europe, Japan, and the United States, and the vast majority work in air-conditioned office complexes. While foot traffic to retail stores has always been somewhat weather-dependent—it's hard to get people into stores when it is raining or freezing cold—these seasonal patterns are already built into the company's annual sales strategy, and if anything the milder winters may help drive a small bump in retail sales.

As with many firms headquartered in New York City, Superstorm Sandy was a bit of a wakeup call. So, in the early 2010s the company hired a contractor to scope out the vulnerability of its assets to coastal flooding; this revealed, to management's relief, that most assets on the company's balance sheet were on high enough ground to be protected from even the worst sea level rise projections, at least for the next 100 years.

A closer look at the company's international supply chain, however, reveals some key vulnerabilities on much shorter time horizons.

Springbok, while responsible for the design and distribution of its branded shoes and apparel, relies on contracts with individual production facilities in many different countries for the manufacture of its products and the various component parts. The bulk of these manufacturers are based in South and Southeast Asia, primarily in India and Vietnam.

So far, heat on the factory floor may not have been front of mind for Springbok's management because it was not their problem to deal with. Occasional heat waves may have put a damper on worker productivity among its Indian and Vietnamese suppliers, but the suppliers were the ones who needed to scramble, hiring extra workhands, and paying employees overtime to fulfill manufacturing orders. In the few exceptional instances when a facility was unable to make a key shipment of shoes on time, Springbok was forced to pivot marketing efforts toward different product lines with greater existing inventory. The loss of revenue was nontrivial, but having worked with these suppliers for many years and given the difficulty of finding a suitable alternate, Springbok chalked up the delay to a freak accident. For the most part, they remained unaware of and did not inquire as to the reasons for the delay.

As the world warms, such freak occurrences are likely to occur with greater frequency. Correlated increases in extreme weather events may increase the likelihood of multiple plants falling behind or failing simultaneously. This could lead to larger potential losses in aggregate, and possibly more systemic threats to the business. Distribution networks may slow due to stoppages caused by increased workplace accidents in warehouses and loading docks, which are expensive to climate control and often not fully air-conditioned even in the developed world. Shipping costs may rise if the wages required to coax dockhands and postal workers to deliver in hotter conditions tend to increase more quickly than average wages.

In developing countries, the confluence of climate-change-driven increases in extreme heat, secular increases in cooling demand due to income growth, and pressures to move away from fossil fuels may put additional strain on electricity grids, leading to more frequent power outages. This may be particularly problematic to firms for whom entire product lines rely on specialized parts made only in a few facilities. For a long time, the same Apple home button was used across the iPhone and iPad product lines, meaning that, even though the cost share associated with the part may have been small, its sudden unavailability would have caused Apple to lose a significant amount of revenue.[7]

As many companies outsource their customer service and basic accounting operations to specialized contractors in countries like China, India, Vietnam, and Thailand, they may also be exposing themselves to greater service disruptions and potential declines in customer service quality due to heat or pollution-related effects on these workers. This could lead to gradual declines in customer satisfaction as heat presses back-office workers even more than it already does.

What I describe above are mostly illustrative hypotheticals. However, empirical evidence on supply-chain propagation increasingly suggests that this kind of risk may be more common than first meets the eye. In a groundbreaking paper using data from 2,051 US-based firms over the period from 1978 to 2013, economists Jean-Noël Barrot and Julien Sauvagnat showed that the propagation of shocks across production networks can be measured in downstream firms' financial performance. They found that natural disasters that affect supplier firms impose large downstream costs on their customers, as measured by changes in market value. Firms whose suppliers are hit by a major natural disaster experienced an average decline in sales growth of two to three percentage points.[8] These effects appear to be most pronounced when supplier firms produce specific inputs for which there are few alternatives, as in the Apple home button example.

In a fascinating paper by Nora Pankratz and Christoph Schiller, episodes of extreme heat and extreme precipitation in countries where upstream suppliers operate led not only to lost production among supplier firms, but also to diminished revenues in downstream companies operating overseas. Moreover, the authors found that, over time, such climate-induced production disruptions can eventually lead buyers to terminate contracts with their suppliers.[9]

An important question therefore is whether investors and management are aware of the effects of such noncatastrophic climate events on companies' bottom lines. Again, while individual cases will vary, evidence suggests that, on average, knowledge of such material impacts may not have been complete even as of recently. While the Barrot and Sauvagnat study found that, on average, natural disaster impacts on supplier firms led to a decline in downstream firms' stock value of around 1 percent, which for larger companies can be in the billions, other recent studies have suggested that the market may not be fully capitalizing the effects of heat-related disruptions in the supply chain. Using data on over 1,800 publicly listed firms' earnings reports, Nora Pankratz, Rob Bauer, and Jeroen Derwall found that while hotter temperatures in the previous quarter adversely affected earnings for a large subset of firms in measurable ways, analysts' expectations did not appear to incorporate this information in their quarterly earnings guidance. In other words, at least in the few decades ending in 2019, financial markets did not appear to have fully capitalized the adverse earnings consequences of *current* episodes of hotter temperature, let alone the *future* changes expected due to climate change.[10]

Recall that, in a study of a wide range of Indian manufacturing firms, hotter temperatures reduced worker productivity substantially, with a 1°C hotter year reducing overall manufacturing productivity by roughly 2 percent. This was not because air-conditioning is not technically feasible in India. In fact, the same study found that, in diamond processing plants, which were mostly air-conditioned, the effects of heat on output were minimal. The important question is whether air-conditioning in such manufacturing plants makes business sense. If relatively low production costs are a primary reason that manufacturing was outsourced to a place like India in the first place, it is unclear how adding air-conditioning in a hot and humid climate would affect the cost-benefit calculus. Other studies found that, for the least-adapted firms, a 40°C (104°F) day essentially wiped out a full workday's worth of output, meaning that a week with such temperatures could reduce annual output by several percentage points.[11]

While it is yet unclear why such effects may not have been fully capitalized, these patterns are consistent with the idea that accurate information about localized physical climate risk may be an important but costly input to climate adaptation. Big companies have an outsized incentive to overcome these information barriers—including the psychological biases that prevent us from seeing climate change for what it is. Perhaps this is why they are already investing heavily in the data analytics necessary to address these biases and the blind spots they can create.

Identifying these vulnerabilities can be done with increasing precision. But to be decision-relevant, they likely must be highly tailored to the risks and potential remedies specific to each firm. Each company may have different potential bottlenecks, each exposed to different vectors of climate hazards. Especially for subtler cumulative risks like hotter temperature, the adverse revenue implications may not become apparent without careful analysis of company data.

This means that such bespoke adaptation intel will likely not come cheap. It also means that more sophisticated, financially well-endowed market participants will be most likely to demand such analytics, giving them an edge over the competition in a changing climate. Indeed, many management consulting companies appear to be betting that climate change will, in addition to creating disruption associated with the clean energy transition, also open new business opportunities in physical risk analytics. In 2021, McKinsey & Company expanded its own climate risk analytics offerings acquiring Vivid Economics, a boutique environment-focused consulting firm. Meanwhile, firms like Jupiter Intel, 427, and Risq have raised hundreds of millions of dollars in funding to offer climate risk data to clients.

We can therefore expect big businesses to find it in their interest to adapt, and to pay for the knowledge and expertise required to do so most effectively. What about for the millions of small to medium sized enterprises and mom and pop stores? Or for households deciding whether and where to buy a home?

We will return shortly to the important issue of whether governments should step in to provide such information or facilitate adaptation and resiliency in other ways. First, let us explore what we know about what physical climate risk may mean for households and individuals and their everyday economic decisions.

Housing Markets and Climate Risk

"In the short run, the market is a voting machine. In the long run, it is a weighing machine." This aphorism is often attributed to British-born economist and investor Benjamin Graham (1894–1976), who many consider to be the father of value investing.

While it is unlikely that this observation was made with climate change in mind, it nevertheless contains a kernel of wisdom instructive for our thinking about climate risk. It speaks to how climate risk is likely to be reflected in—and how its effects may propagate through—financial markets, ultimately affecting households, including those not obviously connected to the physical climate risks in question.

Benjamin Graham was problematizing the superficiality of day-to-day stock price movements, which, he believed, react to salient news events and exhibit significant groupthink, drawing a distinction from what he viewed to be a chief function of financial markets: namely pricing assets properly according to fundamentals, of separating signal from noise.

To extend Graham's metaphor, one could say that an important function of financial markets is to construct accurate weighing systems for anything that can be sold on the marketplace. In modern times, this includes a dizzying array of assets, ranging from shoe companies to ocean-front condos to municipal bonds, not to mention complex derivatives and options stemming from their future income streams. Some of these assets will generate steady income over a long period of time, while others may pay out big or go spectacularly bust. The construction of these metaphorical scales is a diffuse process: an emergent phenomenon that is the result of millions of actors making judgment calls about the risks and potential rewards associated with holding a particular asset.

When the system gets the weights right, the price of an asset—be that a stock, a property, or a bond—should reflect its risk-adjusted, present discounted value. If the asset in question can be analogized to an apple tree, the price reflects how much, in today's dollars, one could expect to make by harvesting and selling the apples it produces during its life, net of the costs associated with planting, watering, harvesting, and so on.

When we say that climate risk—or any risk, for that matter—is fully capitalized in the price of an asset, we are referring to the collective appraisal of the financial market hive-mind determining how such risks should be reflected in the market price. If a company's stocks are traded at a price of $150 per share, the implication is that, given the risks and potential payoffs associated with this asset, the marginal investor values that stock at around $150 per share.

In the case of climate risk, an important question pertains to the transition dynamics between a state of the world in which the risks associated with climate change and its attendant policy responses are not yet reflected in asset prices, and some future state of the world in which most (if not all) such risks are fully capitalized. As climate risks become more salient to investors, price adjustments for more exposed assets are to be expected.

Increasingly, evidence supports the worry that these transition dynamics may not be orderly and smooth. Indeed, some suspect that because of a range of market failures there could be sudden cascading price shocks in response to new information, the ripple effects of which could significantly increase volatility and destabilize financial markets in ways that are systemic.

Bank of England governor Mark Carney calls this the *tragedy of horizons*. A lot of people may have an increasing sense of real future risks, but if most decision-makers do not have a long enough time horizon to do something about it, and the can keeps getting kicked down the road, it becomes increasingly likely that markets may experience a sudden price correction when the reckoning finally occurs. This could apply to transition risk as well as physical climate risk.

Let us consider the case of housing markets. Housing may be one of the most important asset classes when it comes to the possibility of so-called systemic risk. Buying a home is one of the most important financial investments a typical household makes. In the United States, home equity constitutes the largest component of wealth after retirement accounts, accounting for nearly 60 percent of net wealth for the median American. Even for non-homeowners, housing costs often comprise the single largest expenditure item in monthly budgets. This means that price fluctuations in the housing market can have significant impacts on discretionary income and quality of life even for renters.

Because housing is a durable asset that often lasts for many decades, we would expect its value to incorporate both current and future risks, including the risk posed by climate change. In theory, in a perfectly competitive, efficient market, the financial consequences of future climate-related risk should become reflected in a house's price the minute that information about such risks becomes available. In practice, the story seems to be much more complicated.

Before we ask whether housing markets adequately capture *future* climate change risks, it's worth asking a simpler question: How well do housing markets perform in capitalizing *current* physical climate risks? For simplicity, let us focus on flood risks as the relevant hazard category. How well do market prices of homes that lie in historically flood-prone areas reflect the potential damages associated with being in the floodplain? Again, for the moment we are restricting ourselves to the simpler question of whether they incorporate historical flood risk, not the expected increase due to climate change.

Research increasingly suggests that, for many properties, the answer may be not very well at all. Many homebuyers appear to significantly underestimate the potential financial losses associated with flooding of their property. By one account, in the US housing market alone, properties located in the flood plain may be overvalued by a combined $26 billion to $85 billion in US dollars (in 2022 dollars, depending on the discount rate used).[12] Estimating whether an asset is under- or overvalued is of course far from an exact science, and the methods are debated. One approach is to compare the realized *discount* associated with an otherwise identical home located in a floodplain with the hypothetical discount that one would expect in a perfectly efficient market. The latter quantity can be estimated using information on the actuarially fair cost of fully insuring a home against all flood-related damages, or the discounted expected damages associated with flooding over the lifetime of a home in a particular location.

Employing such methods, economists Miyuki Hino and Marshall Burke found that homes in the US floodplain areas (i.e., homes in Special Flood Hazard Areas [SFHAs], which are areas designated by the federal government as having a greater than 1 percent chance of flooding in any given year) sell at a discount of approximately 2.1 percent relative to comparable houses outside the floodplain. In other words, an otherwise identical house that would sell for $500,000 at a safe elevation would sell for $10,500 less if in federally designated floodplain areas. They estimate this by using research methods similar to those discussed in earlier chapters, using data on individual houses and real estate transactions over multiple years and exploiting quasi-experimental variation in floodplain designations, that is, changes in federally determined SFHA boundaries over time.[13]

This suggests that at least some homebuyers are capitalizing flood-related risks into their offer price. But how well does this capture the true financial risks associated with flooding? The −2.1 percent discount appears to be significantly less than what Hino and Burke estimated would be the case if all participants in the housing market were equally well-informed. This *efficient market benchmark* comes out to something more like −4.7 percent, or as high as −10.6 percent if one uses a lower 3 percent discount rate. The gap between the efficient market benchmark and the actual discount suggests that homebuyers may not be adequately incorporating flood risk into their purchasing decisions. These and other recent studies using similarly transparent methodologies found evidence that even current flood risk may not be fully capitalized into housing markets.

Importantly, from a policy perspective it appears that the rate of capitalization varies in perhaps concerning ways. More sophisticated buyers appear more likely to take flood risk into account when making their offers. These include commercial entities that own and rent out multiple single-family homes, as well as family LLCs. They also include households living in areas that have historically been more prone to flooding in general (specifically, counties where more than 10 percent of houses are in the floodplain). Furthermore, recent evidence by Laura Bakkensen and Lint Barrage suggested that low-income and minority residents are more likely to move into high-risk flood zones. This echoes our earlier discussion about who is more likely to be adversely affected by climate change due to differences in the propensity to adapt.[14]

Because of the interlinkages with government programs, which are ultimately taxpayer funded, these housing-related price dynamics have implications for everyone, regardless of whether one lives in more flood-prone parts of the country, or even whether one is a homeowner at all. In many countries, the government acts as an implicit (or explicit) insurer of last resort. In the United States, a large fraction of mortgages is backed by federal loan guarantees in the form of Freddie Mack and Fannie May. In California, private insurers have all but completely retreated from home wildfire protection, leaving the state increasingly on the hook as the provider of such insurance, with an 80 percent increase in the number of policies on the state's books since 2018 alone. Moreover, government-backed insurance schemes often feature premiums that generally do not cover payouts, which means that losses must be covered with general tax revenue or debt issuance.

For these and other reasons, governments are increasingly concerned about the implications for taxpayers and fiscal stability. As a 2023 White House Council of Economic Advisers report warns: "Rapid changes in asset prices or reassessments of the risks in response to a shifting climate could produce volatility and cascading instability in financial markets if not anticipated by regulators."[15]

This discussion illustrates how the propensity to adapt to climate change may vary across individuals and how the resulting price dynamics could subtly affect financial stability. How climate risks are capitalized may shift gradually, or they could shift suddenly. Who can adjust more quickly and who will end up footing the bill is likely to become an increasingly important and heated debate across the world. Evidence increasingly indicates that real estate markets are particularly susceptible to such correlated price shocks that may lead to broader financial contagion.

Not all homes exposed to flood risk are in areas amenable to protection via large-scale environmental engineering, like the Thames River Barrier or the proposed New York City sea wall. Even for those homes that are, questions arise regarding the appropriate policy response, both around whether government intervention in private affairs is warranted and how to choose among a range of potential adaptation options. If one considers the important role that public entities such as Fannie Mae and Freddie Mac play in housing markets, the extent of implicit government involvement is likely nontrivial. As of 2021, over $8 trillion of the $12 trillion in mortgage credit risk in the United States was government-backed, meaning that the issue of adaptation is often intimately intertwined with ongoing debates over the role of government in the economy.

Government Interventions in Climate Adaptation: A Public Economics Framework

As with failures of bodily systems, market failures can come in many forms. Often, multiple market imperfections may overlap, making it more difficult to diagnose the problem and devise the appropriate treatment regimen. In some sense, economic theory and empirical evidence can give us guides to the most common patterns of market failure and how they arise, not unlike how medical sciences help identify common patterns of disease in the body and their causes.

For those who can pay and for goods and services less subject to market imperfections (e.g., home air-conditioning or bespoke climate risk analytics for large corporations), the private market will likely do a good job at providing climate adaptation. Indeed, there are good reasons to believe that, as climate change progresses, markets will come up with many innovative adaptation solutions. However, there will likely be important gaps that the private sector may not fill and situations in which government coordination is necessary. There may also be distressing distributional consequences of leaving adaptation up to the free market.

As with chapter 11 on mitigation, we can appeal to a similar framework of externalized costs to distinguish between areas where private markets will adapt well on their own and where government intervention is likely needed. Indeed, the first fundamental welfare theorem also helps us to identify the situations in which government intervention in adaptation may be most beneficial.

Below is a nonexhaustive list of the major categories of market distortions, imperfections, and failures that may arise in the realm of adapting to climate change.

Incomplete and Asymmetric Information

Incomplete information refers to a situation where participants in a transaction do not have all the relevant information necessary to make an optimal decision. Information asymmetries occur when one party of a transaction has more or better information than another. Information problems outside of the climate change sphere include imperfect information in the context of career or job-related decisions, and information asymmetries between used car sellers and buyers.

It is difficult to manage risks that can't be measured. Irrespective of political stripe or ideology, we can probably all agree that when people are investing their financial resources, they should have easy access to the relevant information.

In the contexts discussed above, it is possible for incomplete and/or asymmetric information to hinder adaptation. For instance, risks in the supply chain may be opaque to managers downstream, or deliberately withheld by suppliers who fear losing clients. In real estate markets, flood risk may be difficult to measure, especially for the average homebuyer. In both cases, the fact that the truth may be costly to discern constitutes a form of market distortion or friction, which can lead to less—and likely less equal—uptake of otherwise socially desirable adaptations.

Increasingly, careful studies of climate risk and economic responses are able to assess the plausibility and potential magnitude of these and other market distortions. For instance, Hino and Burke found that homebuyers appeared more likely to incorporate flood risk into home price offers in states that mandate flood-related disclosures early in the homebuying process. Homebuyers in states with the strictest disclosure laws appear to discount flood risks by nearly twice as much as the national average.

When reliable information regarding climate risk is costly to obtain, government can help facilitate better adaptation decisions by individuals and businesses. This could be by mandating disclosure of some types of climate risk; creating standards for information provision to consumers, as in the case of energy efficiency labels; or investing in the scientific infrastructure necessary to generate more accurate climate forecasts at the local level.

Weather forecasts are another form of information that may be important for adaptation, but which may be underprovided by the free market. Research by Jeffrey Shrader, Laura Bakkensen, and Derek Lemoine suggests that, whether in the context of short run (e.g., daily) weather forecasts, or longer-term predictions regarding El Niño–Southern Oscillation cycles, the value to economic agents of more accurate forecasts can be considerable, whether for construction firms making hiring decisions or individuals protecting themselves from the health impacts of adverse weather conditions. Investing in accurate weather forecasting systems may be especially important for adaptation in countries where, historically, the availability and accuracy of weather forecasts has been relatively poor.[16]

Credit Constraints

Broadly speaking, credit constraints refer to limitations or restrictions on an individual's ability to obtain credit or borrow money from financial institutions. They may arise due to such factors as lack of collateral, information asymmetries between lenders and borrowers, or the lack of accessible banking facilities.

As we saw in chapter 10, adapting to climate change may be hindered by credit constraints. This problem may be particularly acute in the developing world, but it may extend in some cases to richer countries as well. Migration out of harm's way is an obvious adaptation measure that can be affected by lack of liquidity or lack of access to credit.

For some households facing increasing risks due to floods, wildfires, or extreme heat, the upfront capital costs of retrofitting a home or revamping an HVAC system may not be readily available, meaning that credit availability may affect their willingness to undertake such adaptation investments and the overall costs of doing so. Similarly, firms with less cash on hand or higher borrowing costs may be less likely to engage in resiliency-enhancing investments that would otherwise be beneficial net of costs.

Here, it is important to note that having high borrowing costs, perhaps due to a history of not paying bills on time, and not having access to credit are not the same thing (though they are related). This speaks to a subtle but significant distinction between rationales for government intervention rooted primarily in economic efficiency (solving market failures) and those rooted in social equity (helping to improve the lot of the disadvantaged), one that may be easily blurred in many climate adaptation contexts, given the patterns of inequality noted above.

Moral Hazard

Moral hazard refers to the phenomenon where individuals or entities are incentivized to take on greater risks or engage in reckless behavior due to the reduced personal consequences of their actions. In some cases, the best policy intervention may be to reform existing regulations that create moral hazard. For instance, publicly subsidized insurance coverage for properties located in high-risk areas prone to climate-related hazards—coastal regions vulnerable to sea level rise or areas susceptible to wildfires—can create moral hazard to the extent that property owners may be less motivated to undertake mitigation measures or relocate to safer areas because they expect insurance coverage to protect them from significant financial losses.

Similarly, farmers who are insured against losses due to heat or drought may be less likely to invest in resiliency-enhancing crop varieties or farming techniques, especially if crop insurance is publicly subsidized. Evidence suggests that, in the United States, federal crop insurance—which can range as high as 60 percent of the premiums—may be leading food producers to be increasingly maladapted to a warmer climate, illustrating the potential importance of overcoming moral hazard problems in facilitating adaptation.[17]

Labor Market Imperfections

In earlier chapters we explored how climate change may pose a wider range of risks to the world of work than previously appreciated, and that these risks may be growing quickly. This may include the direct health risks of extreme heat, wildfire smoke, and storms, as well as the indirect impacts associated with income volatility arising from a wide range of hazards. Whether or not workers and their employers are able to make the investments and adjustments in practices necessary to mitigate these risks is not immediately obvious. On the one hand, employees and employers may have strong incentives to get adaptation right because failing to do so may directly erode economic competitiveness and livelihoods. On the other hand, it may be that the many market distortions and frictions discussed above create frictions that hinder even mutually beneficial adaptation investments.

For instance, switching a job can be affected by the presence of its own kind of credit constraints because the risk of being temporarily unemployed might pose an insurmountable barrier for those with little savings or limited access to credit to fall back on. This may lock workers into jobs that are increasingly unsafe and unpleasant, even though it would be in their best interest to switch jobs. This could in turn affect workers' willingness to inform their employers of workplace investments or changes in practices that might be worthwhile for everyone given the benefits to workers' health and safety relative to the costs because they fear termination.

Once again, the fundamental question has to do with the extent to which we believe labor markets operate near the competitive market ideal delineated by the first fundamental welfare theorem. Whether and how these factors affect the uptake of technologies and practices that enhance climate resilience at work are open questions. Given the large number of workers potentially exposed to increasing climate risk, and the proportion of these who work in the informal sector, these questions may be of growing societal importance.

Externalities and Local Public Goods

As discussed in previous chapters, externalities arise when there are unintended and uncompensated impacts of an economic activity on third parties who are not directly involved in the activity. Local public goods can be thought of as a special case of positive externality in that they typically involve goods or services that benefit everyone in a relevant community, regardless of whether a given individual paid for it. Public radio broadcasts, the courts, and national defense fit under this category.

In the context of climate adaptation, investments in fireproofing one's own home may provide positive externalities to neighbors, if such investments make it less likely for a wildfire to jump from one home to another. Conversely, in the context of subsidized water provision (which is common in the American West, for instance), growing water-intensive crops or maintaining large lawns may generate negative adaptation externalities. In less water-stressed environments, urban tree cover can provide a range of local benefits, including increased protection from stormwater runoff, reduced heat on hot days, and possibly improved air quality, giving them qualities of a local public good.

In general, when something involves a positive externality or has elements of a public good, the free market is likely to provide too little of it. Conversely, goods and services with negative externalities will be overly abundant. This implies that, to the extent that adaptation and resiliency-enhancing measures have positive externalities, we may want governments to nudge the market in the direction of providing more of them. Similarly, if individuals can choose among a variety of adaptation options, some of which involve more negative externalities than others, it may make sense for lawmakers to discourage those options via a tax or some other form of corrective policy. Figuring out when this is the case, and how much intervention is warranted, is a challenge of growing relevance for public discourse.

Climate Adaptation for the Most Vulnerable

Not all government policies in adaptation need to be motivated by market distortions. Sometimes it may be based on a moral principle, including a desire to improve equity and protect the most vulnerable populations.

In general, wealthier individuals appear better able to prepare for climate-related shocks, recover from their adverse impacts, or find ways to avoid them altogether. This suggests that without additional policy efforts that aim to counteract such forces, poorer and marginalized individuals are likely to bear a disproportionate burden due to climate change—not only because they are more vulnerable to begin with, but also because their rate of adaptation may be slower than others.

The Archbishop Desmond Tutu gives poignant voice to this concern when he speaks of an "adaptation apartheid," where the rich world adapts in the short run, while the poor world, unable to undertake necessary expenditures, faces growing challenges to economic development.

As the archbishop puts it:

> Allowing that drift to continue would be shortsighted. Of course, rich countries can use their vast financial and technological resources to protect themselves against climate change, at least in the short term, that is one of the privileges of wealth. But if climate change destroys livelihoods, displaces people and undermines entire social and economic systems, no country—however rich or powerful— will be immune to the consequences.[18]

Whether or not one agrees that this moral imperative of climate adaptation should be prioritized among other global challenges is a more subjective and nuanced consideration. However, what is clear is that the technical details of what interventions and supports to prioritize, whom to target, and how to implement a given intervention are nontrivial knowledge gaps that require urgent mending.

By now, I hope it has become apparent that there may be more to climate vulnerability than first meets the eye, and consequently many more potential points of effective intervention. Acting on climate adaptation with data-informed strategies can make the difference between a world that adapts quickly and smoothly to coming climate shocks and one that is left reeling and increasingly unequal. There may thus be great value in scientific research that helps clarify when and how government intervention can alleviate adaptation pain points faced by the world's poor.

This is especially true to the extent that some of the most important investments in climate resilience may have little to do with physical technologies. In the previous chapters, we looked at how access to basic banking services may be a critical determinant of whether heat and drought ultimately lead to more lives lost, particularly in agriculture-dependent parts of the developing world. We saw how, for many farmers, the availability and affordability of more drought- and heat-resistant crop varieties may only be part of the story—how overcoming psychological and knowledge barriers may also be key to their effective and widespread adoption. We also considered the role that social norms and local institutions may have in shaping how climate shocks ultimately affect inequalities in educational achievement and attainment.

If one is motivated by a desire to help the world's most vulnerable populations, the coming decades may be critical in fortifying their defenses against the warming that is unavoidable. This is true not only in mobilizing the resources required to help enhance resiliency, but also in generating the knowledge and know-how necessary to deploy those resources most effectively in a changing climate.

Toward an Evidence-Based Adaptation Playbook

Historically, evidence on adaptation has relied on extrapolations of lab-based or engineering estimates of potential adaptive technologies and behaviors to future climate scenarios. Studies have estimated the amount of time workers may lose due to heat by taking ergonomic models of heat transfer based on ambient temperature and exertion levels and then extrapolating the percentage of labor hours lost due to temperature being too hot for sustained human physical activity. Alternatively, studies have relied on engineering estimates of the extent to which sea walls and other technologies may reduce the damage associated with future sea level rise and the resulting increased frequency and intensity of storm surges. Others have estimated the ability of different crop varieties to withstand heat and drought using similar extrapolations.

However, an important limitation of these studies is that engineering estimates often diverge from reality due to factors that models cannot capture. A vast social scientific literature documents the many reasons why realized energy efficiency improvements often fall short of estimates simulated in the lab. Often, engineering estimates also do a poor job of capturing how uptake, adoption, and realized effectiveness of various technologies may vary by socioeconomic factors such as income, gender, and occupation. Simulation models can tell us how much cooler classrooms might be on a hot day if the latest HVAC systems are put in place, but they are limited in telling us why up to 45 percent of classrooms in the United States still do not appear to have adequate air-conditioning or why the distribution of school air-conditioning is highly correlated with race, even conditioned on average local climates.[19] Similarly, process-based methods of estimating agricultural yield responses in a changing climate, while providing detailed analysis of how natural processes may affect yield, often provide only stylized characterizations of how farmers will choose to adapt in the future.[20]

Over the past decade, refinements in our ability to estimate credibly causal *dose–response relationships* between climatic variables such as temperature, precipitation, or floods and economic variables of interest have allowed us to say much more about adaptation in an empirically grounded way. As we discussed previously, studies have increasingly been able to not only establish a causal relationship between changes in temperature and such outcomes as mortality and crime, but also to estimate how these relationships vary with income and average climate.

This is an important development from the standpoint of adaptation because it begins to uncover relationships that inform our understanding of not only the technical potential for human adaptation, but also the expected speed, the realized costs, and the factors within our control that may accelerate adoption in some cases and hinder it in others.

Utilizing similar data and techniques to the study of heat and health described above, the Climate Impact Lab researchers documented evidence of adaptation in the context of energy use and agricultural yield, wherein the slope of the causal relationship between temperature and energy use (as well as that between temperature, precipitation, and agricultural yield) appeared to vary significantly by average income and experience with a given average climate. These findings are in line with other studies, like those by Max Auffhammer in the context of household energy use, that independently find that the way people respond to hotter or colder temperature in terms of energy use varies systematically with income and average climate, which appears to be related to the preponderance of heating and cooling appliances that are suited to a particular climate.[21]

There are some important wrinkles in these studies worth noting. For instance, for much of the world's poor, there is almost no discernible relationship between daily temperature and energy use. This may be because at low income levels, households simply do not have the resources to afford appliances, like air conditioners and refrigerators, that can utilize electricity to blunt the negative effects of extreme temperatures on well-being. It is also worth noting that there are many unresolved puzzles, like why labor supply responses to extreme temperature do not appear to vary as sharply by average climate, or why mental health impacts of hotter temperature appear quite constant across settings.

But the emerging evidence indicates that adaptation can blunt a large share of climate change's adverse human consequences, and with the right data, policymakers may have the ability to facilitate the most effective adaptation investments and target them to those most in need.

Housing Mandates and Wildfire Protection

One example of such an application comes in the context of wildfires. Among the many decisions that communities may face is the thorny issue of how to regulate the construction and operation of buildings that may increasingly be in harm's way, as well as how to insure would-be homeowners against the risk of wildfire-related losses. Should the government provide homeowners monetary assistance ex post to help them cope and rebuild in the wake of fires? Should they incentivize households to move out of increasingly fire prone areas? Or are there alternative options?

Fundamentally, these are questions pertaining both to the proper role of government and the cost-effective design of public policy. In this case, there may be several reasons why the first fundamental welfare theorem doesn't hold. Information imperfections may be at play because it is likely not obvious to the homebuyer what the location-specific wildfire risks are. Nor is it apparent what kinds of materials or practices will be effective at protecting a home from a wildfire's flames, and whether the costs of acquiring these materials and installing them are worth the expected benefits. This is especially true if objective measures of a property's future wildfire risk are not readily available. In addition, there may be potential spillovers and associated externalized costs if one home's installation of fire-resilient materials influences the resiliency of neighboring homes.

In California, mandates were passed in the late 1990s and once again in 2008 that significantly increased the fire resistance of newly constructed homes, for instance by requiring fire-resistant roofing and exterior siding, as well as regulating how decks, building appendages, and covered vents should be built.

Economists Patrick Baylis and Judson Boomhower brought rich housing data to bear on the question of whether and to what extent these policies were not only effective at reducing a home's own fire risk, but also whether the homes that were built to the more stringent standards had positive externalities on neighboring homes.[22]

In addition to putting together a massive data set that contains parcel-level information on when a home was built and whether it was destroyed during a wildfire, they exploited the fact that within a given neighborhood some houses were built (or rebuilt) after new building codes went into effect and some before to ask whether policies that mandated more fire-resilient buildings were protective. The randomized experiment they tried to mimic is one in which otherwise similar homes are exposed to the same intensity of wildfire but only vary in terms of whether they were built before or after the housing codes were put in place. One can imagine that the results of this experiment could be of immense practical value to other jurisdictions weighing the pros and cons of various wildfire mitigation strategies. To control for the fact that other things may have been changing coincidently with the policies (e.g., changes in home-building materials generally), they included data on homes from other states and jurisdictions that were not subject to these building codes and compared the change in building resilience relative to average changes in areas that were not subject to such regulations.

The results were striking. They found that homes built after new codes were significantly less likely to be destroyed during a wildfire. On average, these homes were sixteen percentage points (40 percent) more likely to survive exposure to a wildfire's flames than homes that were built before the new building codes.

Importantly, they found that having one or more neighbors who did not have such fire-resistant materials within ten meters increased the probability of a house being destroyed in a fire by two percentage points (6 percent) per neighbor. This implies that having a home built to these new codes not only confers an own-home benefit, but it also provides a protective benefit to one's neighbors, presumably by making it less likely for flames to jump from one structure to another.

Furthermore, when the researchers computed the implied cost-effectiveness of these interventions, they found that the programs likely have large net benefits—at least for new construction—even at current wildfire risk levels. They noted that the cost of building new homes up to fire-resistant code ranged from between $7,500 and $15,600 per home, and that retrofitting existing homes in this way totaled approximately $62,000 per home. By combining estimates of the cost of reconstruction, alternative living arrangements, and removal of hazardous debris once a house is burned down during a wildfire with zip-code level data on the annual wildfire hazard rate (the probability that homes in a given area are exposed to wildfire in a given year) and comparing them to the costs of building to code, they were able to provide a measure of cost-effectiveness for both new construction and retrofit options.

Their results suggest that, for annual wildfire hazard rates of 0.45 percent or greater (and with conservative assumptions about individuals' preferences for risk, the evolution of future wildfire risk, and other public costs associated with wildfires, including increased taxation for firefighting), these mandates deliver more economic benefits than costs. This is good news for taxpayers, given that, at least in California, many zip codes have wildfire hazard risks above 0.45 percent and climbing, with some zip codes already averaging 2 percent or more. On the other hand, their estimates suggested that mandating full retrofits for existing homes may not be cost-effective, except for the most highly fire-prone areas.

The fact that many home builders in areas with high fire hazard risk were, even very nearly prior to the mandates, choosing not to use such fire-resistant materials and practices despite the overall economic benefits of doing so suggests that there may be important market imperfections at play. This could be due to buyers' and/or builders' perceptions of wildfire risk, moral hazard arising from state and federal support for firefighting and dislocation-related costs, or the positive spillovers whose benefits are being captured by other homes in the neighborhood.

I share this example not to suggest that all parts of the world that are experiencing increased wildfire risk should adopt these sorts of mandated building codes, or that these results imply that building codes should be a prioritized solution among the many alternatives, but rather to illustrate the broader potential for evidence-based adaptation policy that advances the public good. Findings like this help to answer the question of whether the choice of where and how to build one's home is a matter in which government should provide some direction (in wildfire prone areas, the answer seems to be yes) and may ultimately help voters and policymakers choose among a menu of policy options.

For instance, one could imagine fire-prone places like northern Australia or southern France adopting similar measures, or at least probing their potential effectiveness quantitatively by combining the estimates of the protective effect documented above with localized estimates of housing density (which determines the value of the positive spillovers), the rate of new building, and projections of wildfire hazard rates to empirically assess whether such mandates might be cost-effective in their region. In some cases, even the more expensive retrofits may prove to be cost-effective.

Social Services and Protection from Heat

Another example of the importance of public investments in adapting to climate change comes in the context of heat and its impacts on health.

Recent estimates suggest that the slope of the heat–mortality relationship can vary by over an order of magnitude between the richest and poorest places. The effect of a 95°F (35°C) degree day increases mortality in India by nearly thirty times as much as in the United States.

What accounts for such differences? How much is due to private investments that individuals can make, such as home air-conditioning, and how much is due to public investments in better public health infrastructure, better warning systems, and better access to greenspace and cooling centers?

While we are only beginning to uncover the answers to such questions, scrutinizing the United States' experience provides some important clues. Not only does the heat–mortality relationship in the United States vary enormously across places, it has also fallen dramatically over time.

According to a study by Alan Barreca, Karen Clay, Olivier Deschenes, Michael Greenstone, and Joseph Shapiro, between 1959 and 1988 the intensity of the heat-mortality relationship fell by nearly 70 percent. Uptake of air-conditioning may have had something to do with this decline. Using both time and geographic variation in air-conditioning penetration, Barreca and colleagues showed that air-conditioning penetration is highly correlated with changes in the heat-sensitivity of mortality in the United States.[23] This is consistent with other work, including some of my own with Josh Goodman, Jonathan Smith, and Mike Hurwitz, which found variations in school and home air-conditioning penetration to be correlated with differences in the way learning and student performance respond to hotter temperatures. In the context of US high schools, we found that having a fully air-conditioned school may reduce the effect of heat on learning by over 70 percent, and having air-conditioning both at school and at home appears to offset the adverse consequences of heat on learning completely.[24]

However, because in neither of these cases was variation in air-conditioning experimental, we cannot be sure whether it was the change in air-conditioning that caused the improvement in outcomes. While the causal relationship between heat and human outcomes (health in the former, learning in the latter) was arguably credibly identified, the variation in this relationship is not so clearly causal. For instance, it is possible that schools that added air-conditioning also tended to have more funding for better teachers or to plant more trees coincidentally. Similarly, changes in local residential air-conditioning penetration may have been correlated with other factors that affect the heat–health relationship; indeed, the period of most rapid air-conditioning adoption in the United States happens to coincide with a rapid expansion of government health expenditures.

In the case of heat mortality, it appears that access to local healthcare facilities may have also played an important causal role. A careful study by economists Jamie Mullins and Corey White used two sources of quasi-experimental variation to study this question: our familiar variation in daily temperature within a place over time, combined with quasi-experimental variation in the rollout of local community health clinics (CHCs).[25] They exploited the fact that federal legislation—during a period referred to as the "great administrative confusion"—determined funding eligibility for CHCs in such a way that made the realized timing and location of CHCs more or less random.

Mullins and White estimated that the presence of CHCs reduced the heat–mortality relationship by around 15 percent, suggesting that while air-conditioning may yet have been an important contributor to the decline in heat-related deaths in the United States, access to healthcare services also appears to have been instrumental in protecting communities.[26] This is important as we look to the future, especially in parts of the world that currently lack both widespread air-conditioning and robust healthcare systems, and are set to experience many more potentially deadly heat events. The important methodological point to note here is that the researchers combined two sources of quasi-random variation to test a highly policy-relevant hypothesis, namely whether changes in access to healthcare causally affect the causal relationship between temperature and health. One can imagine similar methods yielding invaluable information regarding the efficacy of various proposed public adaptation interventions, either using quasi-experimental variation or by piloting programs that can harness the power of randomization.

Alan Barreca, Paul Stainier, and I found evidence of a similarly important public role in blunting the adverse consequences of heat, especially for poorer households. Recall the study that measured the effects of hot days on the risk of electricity disconnection for low-income Californian households. Many low-income households in the United States already receive some modest amount of public support to help pay their energy bills. Importantly from the perspective of causal inference, there are quirks in the eligibility for such programs that allow the comparison of otherwise similar households just above and just below eligibility cut-offs, which helps explore whether having access to such means-tested subsidies is protective. When we assessed how the heat–disconnection relationship varied by a household's eligibility to receive energy assistance, we found that there was a sharp and discontinuous difference in disconnection risk associated with being ineligible for and thus not enrolled in the program.[27]

What do findings such as these imply for public policy? On the one hand, they provide further proof of concept for how data may inform the design of evidence-based adaptation policy. Rather than simply speculating what might be protective based on engineering estimates or transferring funds to local communities without coordination or guidance regarding best practices, policymakers can in principle prioritize adaptation investments based on what has been proven to work. Of course, given the hyper local nature of many adaptations, it will likely be important to assess the applicability of the findings and to engage local communities early and often regarding their specific adaptation challenges. On the other hand, they speak to important knowledge gaps that need urgent filling. While evidence increasingly suggests that a range of adaptation solutions—whether it is air-conditioning or CHCs—may be effective, in settings where policymakers must make trade-offs amid resource constraints, additional information may be necessary to discern not only the cost-effectiveness of a given intervention but also its context appropriateness.

Suppose a country or a city government is interested in reducing the health burden of hotter temperatures. One can imagine several ways of achieving this objective. One might be to simply invest pell-mell in growing the economy, with the hope that higher incomes will buy everyone better protection against extreme heat. Another might be to target income supports by taxing the broader population and providing cash transfers to the lowest-income households. A third might be to provide means-tested energy subsidies for households that might not yet have reliable access to cooling, like home air-conditioning or heat pumps. A fourth alternative might be for the government to invest in a series of programs, including local community health clinics, but that would likely not be possible without public coordination. Of course, in an ideal world, one would not have to choose among these options. But in the context of limited resources, it would make sense to choose the ones that maximized social returns, especially if such expenditures required raising taxes.

Measure Twice, Cut Once:
Bigger Data and Better Targeting

Another area where important progress is occurring is in targeting the most disadvantaged individuals and communities. As countries scrambled to prevent job losses and a complete economic collapse in the wake of the COVID-19 pandemic, they resorted to policy instruments like the Paycheck Protection Program in the United States. While the Paycheck Protection Program did manage to save many jobs, its lack of precise targeting led to a high total cost to government, and ultimately to taxpayers.

According to studies by David Autor and a team of economists, for every dollar in wages saved by the Paycheck Protection Program, $3.13 went to other areas. They estimated the cost of saving one job for a year to be $169,300, much higher than the average compensation for those jobs, which was $58,200. This suggests that the Paycheck Protection Program's success in preventing job losses during the pandemic came at a high fiscal cost, highlighting the importance of precise targeting and efficient allocation of resources in policy design.[28]

As we have seen, climate change will pose a challenge to society not only because of its disparate impacts on rich and poor countries, but also because its impacts array differently across richer and poorer people. This begs the question of how to better target those most in need, particularly in data-poor settings where figuring out whom within a region or a community may be most vulnerable is not an easy task. Once we know what the adaptation market imperfections are and what kind of government interventions to administer, we may improve cost-effectiveness through better targeting. Here, advances in our ability to target poorer households may help.

How do you target the poorest households when there are few reliable public records of household income or wealth? These challenges are especially acute in developing countries, where data collection infrastructure is limited and where local bureaucrats may have an incentive to misrepresent certain statistics. Some estimates suggest that fewer than half of all countries have reliable access to data on poverty, income, or wealth at levels that are disaggregated below the largest administrative level (e.g., state) or not subject to political censorship.

Advances in remote sensing, combined with adroit uses of big data, including social media data, provide important opportunities to supplement existing measures of household wealth.

For instance, a team of researchers led by Guanghua Chi have put together a fascinating data set that provides micro estimates of poverty and wealth at the 2.4-by-2.4-kilometer level across all 135 low- and middle-income countries. This includes average absolute wealth estimates as well as measures relative to others in the same country, measures that have typically been hard to generate comprehensively and with high geographic resolution.[29]

How do they do this? They first compile a data set composed of detailed satellite images, mobile phone network data, and Facebook connectivity data spanning a wide range of countries and geographies. They then combine this with far more detailed in-person survey data on a specific subset of communities using data compiled by the US Agency for International Development's Demographic and Health Surveys. Using machine learning to convert the raw data into a set of quantitative features that can predict local wealth based on available features with a high degree of accuracy, they created a data set that describes predicted local wealth across all 135 low- and middle-income countries for which the satellite, mobile phone, and Facebook data are available.

The predicted data can explain about two-thirds (56 to 70 percent) of the actual variation in material wealth as described in the detailed in-person surveys, which asks very specific questions regarding the kinds of assets (e.g., furniture, livestock) that households own, the kind of work members of the household engage in, and how much land they own.

Such innovations open the possibility of more cost-effective targeting of adaptation assistance than many previous data collection methods.

From Silver Bullets to Silver Linings

Will societies adapt to the successfully blunted but significant climate change that is in store in the twenty-first century? This chapter has highlighted the need for more nuanced and holistic approaches to adaptation, which consider not just technological fixes, but also the social and economic dimensions of climate change.

The pervasiveness of climate change's more subtle impacts necessitates an equally holistic approach to thinking about climate adaptation—one that takes seriously the need not only for grand public works, but also for the mundane, often messier adaptations to the way society organizes its means of production and ensures against personal calamity for its most vulnerable.

Understanding how to cost-effectively and equitably adapt to climate change is an increasingly important question for which we currently only have partial answers. Given the way markets work, we have good reason to believe that financially well-endowed participants in the private sector will seek ways to protect themselves. In the case of real estate markets and large corporations with international supply chains, for instance, more sophisticated actors are already incorporating new information regarding slower burning climate risks into their decision-making. Some are actively investing in processes that help them uncover the information required to make more climate-resilient decisions by hiring the services of boutique climate risk consulting firms.

The growing abundance of private climate risk analytics will help many firms and investors adapt to climate change. But for most businesses and individuals, such tailored information services may be out of reach. Indeed, many of the climate risks outlined in this book may not be very salient, not rising to the top of a crowded list of issues vying for precious mental bandwidth and limited financial resources. This does not mean that they will not adapt. Evidence suggests that in general and over time humanity can be remarkably adaptive. However, it does mean that some risks may be easier to adapt to than others, for some more quickly than others, and that figuring out the collective constraints to effective adaptation may be among the most important climate action items in the coming decades.

For most individuals and businesses, adapting to climate change will likely benefit from some form of public coordination and facilitation, especially in cases where adaptation is hindered by market imperfections. This is not the same thing as government financial support, full stop. In fact, an important framework introduced in this chapter is a public economic approach to discerning the *when* and *how* of government intervention in adaptation, one that is rooted in the first fundamental welfare theorem of economics.

We explored various forms of market imperfections, including information asymmetries, local externalities and public goods, moral hazard, and credit constraints. While more research is urgently needed to better test the extent to which these constraints bind and for whom, we saw how careful empirical analyses of earlier policies, like the rollout of CHCs or fireproofing building mandates in California, might help policymakers make better decisions about adaptation. There may also be value in more transparent and principled modeling of local climate projections, especially as they make their way into consumer-facing products, such as Redfin's flood score function for real estate listings—a form of information public good that governments can help to coordinate.

Given the interplay between climate change and global economic inequality, considerations of equity and fairness may feature heavily in determining how tax revenues are spent in the service of adapting to a warmer climate. Here, too, better use of data can make a big difference, both in identifying the factors that make some populations more vulnerable, and in better targeting the households and communities that are most likely to be in desperate need of assistance. In some instances, reforming existing government policies, including social safety net programs such as public lending, insurance, and welfare so that they are updated to reflect the challenges of a changing climate may be the first port of call.

Without learning about the day-to-day details of how noncatastrophic climate change hurts, we cannot hope to design effective adaptation solutions. Without recognizing the diffuse but cumulatively important nature of the slow burn, we might risk focusing on the wrong climate adaptation policies.

As we have seen, some of the most important adaptations may have little to do with silver-bullet technical solutions, but rather comprise a messy mishmash of incremental adjustments that help make our social and economic systems more resilient to the various risks that climate change brings.

Conclusion

THE REST OF CREATION

GROWING UP, I had the blessing of spending copious amounts of time outdoors. Mother Nature was my playground; her innumerable fauna and flora were my friends. Perhaps that is why, for much of my childhood, I wanted to be a biologist when I grew up.

I will never forget the feeling of waking up to the chorus of tree frogs and sulfur crested cockatoos in the rainforests of Northern Australia. As part of a semester abroad with the School for Field Studies, a dozen or so college students and I spent the spring of 2005 living in a stilted cabin in the middle of the rainforest, studying the plants and animals around us and exploring the many nooks and crannies of a dense tropical paradise. The sight of a seven-foot monitor lizard rummaging through our garbage is still vivid in my recollection, as are the few rare sightings of tree kangaroos high in the rainforest canopy.

One experience during my time in Australia stands out. This was when I had the opportunity to go snorkeling on the Great Barrier Reef. Most of my exposure to coral reefs had been through books and documentaries. I knew enough to be curious. But I had no idea what I was in for.

We had taken a small catamaran out of Cairns, a coastal town a few hours north of Sydney. Not accustomed to the constant rocking of a smaller boat, I became quite seasick, and retired below deck, in my mind having thrown in the towel on sightseeing for the day. Thankfully, the skipper, who was a stocky Aboriginal man, dragged me up, shoved a snorkel in my mouth, and forced me off the deck and into the ocean, yelling as he launched me overboard: "you'll feel better once you're in the water!"

He couldn't have been more on point. After the initial shock of entering the water passed, I became mesmerized by a kaleidoscope of color and light the likes of which I had never seen before. All memory of seasickness vanished from body and mind.

The vibrancy and diversity of life was astounding: fan-shaped corals flanked by schools of tiny, glimmering fish; giant clams the size of a human torso; hundreds of sea urchins and sea anemones and the tiny clown fish they protected. There was so much to marvel at, so much intricate detail. In the mind's eye, there was grace in the movements of the creatures inhabiting this underwater citadel. The delicate dance of the seahorse, the intelligence and curiosity of the octopus, the equipoise of a passing sea turtle. Minutes before, this trip couldn't end soon enough. Now, I was in a race against time to take in every detail.

Perhaps the most awe-inspiring aspect to me was the harmony of it all. Here was a collection of thousands of living creatures so varied in shape and form and behavior, somehow arranged in a symbiotic system that had evolved over millennia to maintain a beautiful, if delicate, balance.

I remember writing to my parents soon thereafter: please add the Great Barrier Reef to your bucket list. And please get around to it sooner rather than later.

For nature lovers like me, these memories—and the thought that such monitor lizards and tree kangaroos still roam the forests of Australia—hold real value, even though the last time I saw either in person was decades ago.[1] I have tried to argue in this book that, irrespective of your personal empathy for the destruction of nature, you should care about climate change because of the economic harm it may do to us, our children, and the other 8 billion human inhabitants of the earth. I have presented evidence, mostly derived from data and statistics, that even based on the noncatastrophic costs of climate change for human beings, there is an economic rationale for aggressive climate action. Implicit in my exposition has been the idea that regardless of what our subjective experiences with nature may be, statistical reality is that climate change has measurable effects on human society, and that the damages they impose may be large enough to merit many of the costs associated with reducing greenhouse gases.

If your value set includes any sympathy for the plight of those who will suffer through no real fault of their own, including nonhuman sentient beings, then the case is arguably even stronger. However, it is worth being clear about the fact that, in talking about my personal experience on the Great Barrier Reef, I am making a more explicit appeal to personal values (namely, my appreciation and respect for the natural world), which means that I have less of an expectation that you agree with or empathize with the way I attach values to the experiences I have had, or how they affect my understanding of climate change.

Of course, one could make an appeal to the practical economic value of reef ecosystems. In addition to their awe-inspiring beauty, coral reefs have outsized ecological significance. Though they cover only 1 percent of the ocean floor, they are estimated to house more than one-quarter of the species in the ocean.[2] Many serve as nurseries for fish species who reside in the open ocean, thus comprising essential links in the complex ecological chains that sustain many commercial fisheries around the world.

Either way, if there is a place in your heart that pangs at the destruction of the natural world, at the loss of beautiful and biodiverse ecosystems like the Great Barrier Reef, then the case to curtail climate change may be all the more urgent indeed, insofar as the risk here may in fact be truly existential.

Or is it really?

You're probably familiar with the idea that coral reefs are threatened by climate change. And if you're anything like me, it probably makes you a bit depressed.

Most visibly, corals are directly affected by warming oceans, which can trigger bleaching, a phenomenon whereby corals eject their symbiotic algae, and which if prolonged can lead to mass die-offs. This is one of the reasons why organizations like the International Union for Conservation of Nature and the UNESCO World Heritage Convention warn that "limiting global average temperature to well below 2°C above preindustrial levels and pursuing efforts to limit the temperature increase to 1.5°C, in line with the Paris Agreement on climate change, provides the only chance for the survival of coral reefs globally."[3]

To be honest, the heuristic I applied in my own mind when it came to climate change and coral reefs was one of catastrophic fatalism, at least for a time. The images of once colorful ecosystems blanched ghostly white seemingly overnight were shocking. My heart was only too eager to dip into that mix of empathic distress and outright despair that such images of ecological destruction can elicit. My mind, too quick to conclude that, at least on this special dimension, the issue was indeed black and white. After all, the mind reasoned, is there not a clear scientific threshold for the ocean temperature above which coral bleaching occurs? And are we not on course to bulldoze through that threshold even with aggressive mitigation? Coral reefs, sad as it is to accept, must be doomed.

The reality, it turns out, may be subtler than that. While there is no disputing the dire circumstances facing most of the world's coral reefs, the evidence suggests that there may also be a significant silver lining that a doomsday disposition may obstruct.

First, some more bad news. Warmer ocean temperature is not the only factor that adversely affects corals. The health of coral reef systems is also affected by ocean acidification. Acidification is another form of the hidden, slow-burning threat in that its effects are very gradual and hard to spot. Unlike bleaching and damaging storms, which episodically mark their presence in more visible ways, ocean acidification is what Kenneth Anthony of the Australian Institute of Marine Science calls a *creeping stressor*, one that "gradually compromises a suite of biological functions on coral reefs." In something of a mirror image to the way carbon fertilization of the atmosphere speeds up photosynthesis of plants, acidification of the ocean slows the rate of coral growth, leading to more fragile structures and slower recovery rates between disturbances.[4]

Neither of these stressors—the warming of the oceans or their acidification—can be properly mitigated without first turning down the dial on greenhouse gas emissions globally. But the health of coral reefs turns out to be affected by a wide range of other factors, some of which may be much more immediately actionable locally. Among them are overfishing in neighboring fisheries, pollution from coastal urban developments, degradation due to unsustainable forms of reef tourism, and runoff of sediment, nutrients, and various pesticides from nearby agricultural land. Research increasingly suggests that, in many coral reef regions, such local stressors are compounding the strain created by warming and acidifying oceans, further reducing reefs' resilience in the face of increasing climate shocks.

Rather than viewing the problem of climate change's impacts on coral reefs through a binary extinction-versus-survival lens, perhaps we may recognize that here, too, the relevant heuristic is one of many shades of gray; one that recognizes reef resilience to be a continuous variable that can erode or improve (or, at minimum, erode less quickly) depending on many human factors. Much like the social and economic aspects of climate mitigation and climate resilience explored in this book, the ecological dimension of climate change may also benefit from thinking statistically in this way.

Unless you are a perennial pessimist, this pivot of perspective offers an unmistakable cause for hope. It suggests that, even in the context of coral reefs, there are many actions that can be taken to improve reef resilience in the face of a changing climate. This may take the form of establishing and expanding networks of marine park areas and no-take fishing reserves, or implementing land use practices that make watersheds less exposed to soil erosion following floods, thus reducing pollutant runoff into coral reef waters. It could take the form of engineering solutions that have not yet been perfected, perhaps in the form of *assisted evolution* whereby corals are assisted in adapting to changing ocean conditions at an epigenetic level; or alkalinity enhancement, which could help buffer against increased acidity in localized habitats. Or it could be as simple as practicing better reef stewardship next time we go to visit, being careful to look but not touch, and to support those businesses that engage in sustainable tourism practices.

The curiosity that such a perspective often leads to can itself be a counter to the darkness of climate fatalism and at minimum give us more avenues through which to channel our concern.

Of course, such hope must no doubt be tempered by realism. There is no sugar coating the fact that, in the case of coral reefs, such resiliency enhancing measures are inherently limited by the physical constraints imposed by warming and acidification.[5] But it seems important not to lose sight of the fact that, especially during the period between now and a stabilized future climate, a great many beneficial local actions can be taken.

If the doomsday climate narrative invokes fear and fatalism, the non-catastrophic slow burn perspective invites us to view climate change with sober resolve, compassion for those most vulnerable, and a sense of active hope.

I have tried to make the case that with a more intellectually curious perspective we may recognize that there are a multitude of ways one can make a difference on climate change, not only by recycling or installing solar panels or shaming politicians into stronger national mitigation pledges but also by helping to make the pain of climate change more manageable for ourselves and the most disadvantaged among us. That the climate change story is one in which every little bit of emissions reduction helps—even if we are too late to prevent the crossing of symbolic temperature thresholds, whether that be 1.5°C or 2.0°C—and one in which human actions often at the local level can make a big difference in how much damage is done as we muddle our way through to a more stable climate.

I believe this sense of hope and agency is perhaps most important for today's young people. Both the young of age and the young at heart need it, as does anyone, really, who is open to new ways of thinking about the world and their place in it.

NOTES

Introduction

1. E. Zanella, L. Gurioli, M. T. Pareschi, and R. Lanza, "Influences of Urban Fabric on Pyroclastic Density Currents at Pompeii (Italy): Part II: Temperature of the Deposits and Hazard Implications," *Journal of Geophysical Research* 112, no. 112: B05214, https://doi.org/10.1029/2006JB004775.

2. G. Mastrolorenzo, P. Petrone, L. Pappalardo, and F. M. Guarino, "Lethal Thermal Impact at Periphery of Pyroclastic Surges: Evidences at Pompeii," *PloS ONE* 5, no. 6 (2010): e11127, https://doi.org/10.1371/journal.pone.0011127.

3. Willem Jongman, "Gibbon Was Right: The Decline and Fall of the Roman Economy," in *Crises and the Roman Empire: Proceedings of the Seventh Workshop of the International Network Impact of Empire, Nijmegen, June 2–24, 2006,* edited by Olivier Hekster, Gerda de Kleijn, and Daniëlle Slootjes (Leiden: Brill, 2007), 183–199.

4. Edward Gibbon and Henry Hart Milman, *The Decline and Fall of the Roman Empire*, vol. 3 (New York: Modern Library, 2003); James W. Ermatinger, *The Decline and Fall of the Roman Empire* (Westport, CT: Greenwood Press, 2004).

5. Economic data from that period is scarce, but evidence from ice cores, shipwrecks, and dated wood remains certainly corroborates the idea that the height of the Roman Empire corresponded with an unprecedented level of maritime trade, metal extraction, and building activity, which subsequently declined significantly. We also see hints of declining standards of living in the deterioration of diet quality suggested by average human bone sizes from the period, and the size and number of livestock such as pigs, cows, sheep, and horses suggested by the rate and composition of animal bone deposition. Anthony King, "Diet in the Roman World: A Regional Inter-site Comparison of the Mammal Bones," *Journal of Roman Archaeology* 12, no. 1 (1999): 168–202.

6. Simon Dietz, Frederick van der Ploeg, Armon Rezai, and Frank Venmans, "Are Economists Getting Climate Dynamics Right and Does It Matter?," *Journal of the Association of Environmental and Resource Economists* 8, no. 5 (2021): 895–921, https://doi.org/10.1086/713977.

7. Prior to the Paris Accords of 2015, median warming projections were 3.6°C by 2100. Inclusive of the Glasgow Pledges of 2021, median end-of-century warming projections stood at 2.4°C.

8. A United Nations report estimates that humanity will "likely need to cut greenhouse gas emissions by an additional 15 to 30 billion tons by 2030" to avert the worst impacts of climate change. Analyses suggest that the US Inflation Reduction Act would supply about 1 billion tons of additional cuts. Nadja Popovich and Brad Plumer, "How the New Climate Bill Will Reduce Emissions," *New York Times*, August 12, 2022, https://www.nytimes.com/interactive/2022/08/02/climate/manchin-deal-emissions-cuts.html.

284 NOTES TO CHAPTER 1

9. Thomas Piketty, Li Yang, and Gabriel Zucman, "Capital Accumulation, Private Property, and Rising Inequality in China, 1978–2015," *American Economic Review* 109, no. 7 (2019): 2469–2496, https://doi.org/10.1257/aer.20170973.

10. While post-pandemic wage and employment growth in some low-end occupations has offset some of the erosion at the bottom, at least temporarily, the fact of a broader trend toward economic bifurcation remains. Such "polarization" in economic outcomes has been playing out in many countries to varying degrees. See, for instance, Maarten Goos, Alan Manning, and Anna Salomons, "Job Polarization in Europe," *American Economic Review* 99, no. 2 (2009): 58–63, http://www.aeaweb.org/articles.php?doi=10.1257/aer.99.2.58; and Piketty, Yang, and Zucman, "Capital Accumulation."

11. FM Global, "CEO/CFO Climate Risk Survey Results," July 2, 2020, https://www.fmglobal.mobi/insights-and-impacts/2020/climate-risk-survey-results. Specifically, that "addressing climate risk is a medium to high business priority."

12. FM Global, "CEO/CFO Climate Risk Survey Results."

13. Michael Greenstone, "Making the Case for Climate Action: The Growing Risks and Costs of Inaction," Energy Policy Institute at the University of Chicago, April 15, 2021, https://epic.uchicago.edu/insights/making-the-case-for-climate-action-the-growing-risks-and-costs-of-inaction/.

Chapter One

1. Niholas D. Kristof and Sheryl WuDunn, *Half the Sky: Turning Oppression into Opportunity for Women Worldwide* (New York: Alfred A. Knopf, 2009).

2. Douglas Broom, "As the UK Publishes Its First Census of Women Killed by Men, Here's a Global Look at the Problem," World Economic Forum, November 25, 2020, https://www.weforum.org/agenda/2020/11/violence-against-women-femicide-census/.

3. Kristof and WuDunn, *Half the Sky*, xiv–xv.

4. Jennifer Marlon et al., "Yale Climate Opinion Maps 2021," February 23, 2022, https://climatecommunication.yale.edu/visualizations-data/ycom-us/.

5. This is according to the Lowy Institute, a think tank based in Sydney, Australia. Lowy Institute, "Attitudes to Global Warming," Lowy Institute Poll, accessed August 7, 2022, https://poll.lowyinstitute.org/charts/attitudes-to-global-warming.

6. For instance, in Russell County, Kansas, the Yale model, which takes a nationally representative sample and projects locally based on demographic factors, estimates it to be 30 percent. Yale Program on Climate Change Communication, https://climatecommunication.yale.edu/visualizations-data/ycom-us/.

7. Jonathan Weisman and Jazmine Ulloa, "As the Planet Cooks, Climate Stalls as a Political Issue," *New York Times*, July 18, 2022, https://www.nytimes.com/2022/07/17/us/politics/climate-change-manchin-biden.html?referringSource=articleShare.

8. Daniel Kahneman, *Thinking, Fast and Slow* (New York: Random House Audio, 2011), part 1, chapter 1.

9. And while there is as yet little scientific evidence to substantiate this, many of humanity's greatest minds pay respectful homage to the ingenuity and creativity dormant in the undercurrents of human intuition. It is perhaps what Ralph Waldo Emerson refers to as touching the

"Soul of the Whole," or Jung's perceptive unconscious. Einstein puts it provocatively: "All great achievements of science must start from intuitive knowledge."

10. Bureau of Labor Statistics, "Injuries, Illnesses, and Fatalities," accessed June 2022, https://www.bls.gov/iif/oshcfoi1.htm#2020.

11. Kahneman, *Thinking, Fast and Slow*, part 1, chapter 1.

12. For instance, Martin L. Weitzman, "The Geoengineered Planet," accessed January 2022, https://scholar.harvard.edu/files/weitzman/files/the_geoengineered_planet_chapter_10.pdf; and Weitzman, "A Voting Architecture for the Governance of Free-Driver Externalities, with Application to Geoengineering," *Scandinavian Journal of Economics* 117, no. 4 (2015): 1049–1068, https://scholar.harvard.edu/files/weitzman/files/weitzman-2015-the_scandinavian_journal_of_economics.pdf.

13. Sea level rise will most certainly cause an enormous amount of damage and likely displace many millions of people. But it is worth noting that what is meant by "eventually" is often unclear or obscured, even though the time frame likely matters nontrivially. Current projections involve global mean sea level rise on the order of 0.5 to 1 meter by 2100. This is a lot, but a far cry from the "most of Miami underwater" scenarios that typically assume several meters on the basis of multi-hundred-year projections that include potential melting of continental ice sheets. For instance, one study assesses the impact of the melting of the Greenland ice sheet on sea level rise over a 1,000-year period. Jonathan M. Gregory, Philippe Huybrechts, and Sarah C. B. Raper, "Threatened Loss of the Greenland Ice-Sheet," *Nature* 428, no. 616 (2004), https://doi.org/10.1038/428616a.

14. Perhaps this is why, even though over 70 percent of Americans believe that "climate change is happening," only 47 percent say that they believe "climate change will harm me personally" and only 55 percent say that it should be a "high priority for the President and Congress." Marlon et al., "Yale Climate Opinion Maps 2021."

15. A subtler variation of this problem arises from the fact that the earth's atmosphere is a highly complex and nonlinear system, which means that climate change sometimes creates weird weather that seems at odds with the notion of warming.

16. In 2015, Senator James Inhofe from Oklahoma brought in a snowball to Congress as proof against global warming; after a cold snap in New South Wales, Senator Matt Canavan of Australia tweeted a photo of the scene with a sarcastic caption reading, "climate change." Ted Barrett, "Inhofe Brings Snowball on Senate Floor as Evidence Globe Is Not Warming," CNN, February 27, 2015, https://www.cnn.com/2015/02/26/politics/james-inhofe-snowball-climate-change; Nerilie Abram, Martin de Kauwe, and Sarah Perkins-Kirkpatrick, "Matt Canavan Suggested the Cold Snap Means Global Warming Isn't Real. We Bust This and 2 Other Climate Myths," The Conversation, June 11, 2021, https://theconversation.com/matt-canavan-suggested-the-cold-snap-means-global-warming-isnt-real-we-bust-this-and-2-other-climate-myths-130878.

Chapter Two

1. "Costliest US Tropical Cyclones," National Centers for Environmental Information, August 8, 2023, https://www.nhc.noaa.gov/news/UpdatedCostliest.pdf.

2. Rajashri Chakrabarti and Max Livingston, "The Impact of Superstorm Sandy on New York City School Closures and Attendance," Liberty Street Economics, December 19, 2012,

https://libertystreeteconomics.newyorkfed.org/2012/12/the-impact-of-superstorm-sandy-on-new-york-city-school-closures-and-attendance/.

3. Lesli A. Maxwell, "Hurricane Sandy Shutters Thousands of Schools," *Education Week*, October 29, 2012, https://www.edweek.org/leadership/hurricane-sandy-shutters-thousands-of-schools/2012/10.

4. Chakrabarti and Livingston, "Impact of Superstorm Sandy."

5. The self-proclaimed "scorekeeper" of climate and severe weather events, the National Centers for Environmental Information (NCEI), notes that the potential damages to learning or human capital more broadly are not counted in these or any other estimates of the economic costs of natural disasters. As NCEI notes: "These costs include: physical damage to residential, commercial, and municipal buildings; material assets (content) within buildings; time element losses such as business interruption or loss of living quarters; damage to vehicles and boats; public assets including roads, bridges, levees; electrical infrastructure and offshore energy platforms; agricultural assets including crops, livestock, and commercial timber; and wildfire suppression costs, among others. However, these disaster costs do not take into account losses to: natural capital or environmental degradation, mental or physical healthcare related costs, the value of statistical life, or supply chain, continent business interruption costs." "Billion-Dollar Disasters: Calculating the Costs: 7 Things to Know about NCEI's US Billion-Dollar Disasters Data," NCEI, August 2023, https://www.ncei.noaa.gov/access/monitoring/dyk/billions-calculations.

6. Sacerdote provides a compelling analysis of the short and long-run educational implications of Hurricane Katrina. Bruce Sacerdote, "When the Saints Go Marching Out: Long-Term Outcomes for Student Evacuees from Hurricanes Katrina and Rita," *American Economic Journal: Applied Economics* 4, no. 1 (2012): 109–135. The effect sizes in Sacerdote are smaller than but in the same order of magnitude as charter school effects. For instance, Atila Abdulkadiroğlu et al. found that winning a charter school lottery in Boston raises ELA test scores by 0.20 standard deviations and math test scores by 0.40 standard deviations. Abdulkadiroğlu et al., "Accountability and Flexibility in Public Schools: Evidence from Boston's Charters and Pilots," *Quarterly Journal of Economics* 126, no. 2 (2011): 699–748, https://academic.oup.com/qje/article/126/2/699/1871552.

7. C. Kirabo Jackson and Claire Mackevicius, "The Distribution of School Spending Impacts" (working paper no. 28517, National Bureau of Economic Research, 2021), https://www.nber.org/papers/w28517.

8. Tatyana Deryugina and David Molitor, "Does When You Die Depend on Where You Live? Evidence from Hurricane Katrina" (working paper no. 24822, National Bureau of Economic Research, 2018), https://www.nber.org/papers/w24822.

9. Indeed, Katrina is very much an outlier in terms of the magnitude of the dislocation, as well as the amount of funding provided to help support such out-migrations. Moreover, many New Orleans public schools were extremely low-performing to begin with, making it much more likely that students moved to better-performing districts.

10. "Costliest US Tropical Cyclones," National Centers for Environmental Information, August 8, 2023, https://www.ncei.noaa.gov/access/billions/dcmi.pdf.

11. Isaac M. Opper, R. Jisung Park, and Lucas Husted, "The Effect of Natural Disasters on Human Capital in the United States," *Nature Human Behaviour* 7 (2023): 1442–1453.

NOTES TO CHAPTER 3   287

12. Morley Gunderson and Philip Oreopolous, "Returns to Education in Developed Countries," in *The Economics of Education*, ed. Stephen Bradley and Colin Green (London: Elsevier, 2020), 39–51.

13. Raj Chetty, John N. Friedman, and Jonah E. Rockoff, "Measuring the Impacts of Teachers II: Teacher Value-Added and Student Outcomes in Adulthood," *American Economic Review* 104, no. 9 (September 2014): 2633–2679, https://www.aeaweb.org/articles?id=10.1257/aer.104.9.2633.

14. Flavio Cunha and James Heckman, "The Technology of Skill Formation," *American Economic Review* 97, no. 2 (2007): 31–47; James J. Heckman, "Skill Formation and the Economics of Investing in Disadvantaged Children," *Science* 312, no. 5782 (2006): 1900–1902.

15. For instance, a study by Nicholas J. Sanders finds that US high schoolers who experienced more prenatal pollution in the early 1980s (due to variation in how the industrial recession affected local factories) score significantly lower on standardized tests in "What Doesn't Kill You Makes You Weaker: Prenatal Pollution Exposure and Educational Outcomes," *Journal of Human Resources* 47, no. 3 (2012): 826–850. A one-standard-deviation increase in county level pollution is associated with a 2%–6% of a standard deviation reduction in test scores. Similarly, a study comparing siblings who were born under different local environmental conditions find that increased local carbon monoxide exposure in-utero leads to lower fourth grade test scores, as discussed in Prashant Bharadwaj et al., "Gray Matters: Fetal Pollution Exposure and Human Capital Formation," *Journal of the Association of Environmental and Resource Economists* 4, no. 2 (2017): 505–542. In fact, Adam Isen, Maya Rossin-Slater, and W. Reed Walker found, using IRS tax data, that higher pollution levels in the year of birth are associated with lower labor force participation and lower earnings at age 30. Isen, Adam, Rossin-Slater, and Walker. "Every Breath You Take—Every Dollar You'll Make: The Long-Term Consequences of the Clean Air Act of 1970," *Journal of Political Economy* 125, no. 3 (2017): 848–902; Janet Currie et al., "What Do We Know about Short- and Long-Term Effects of Early-Life Exposure to Pollution?," *Annual Review of Resource Economics* 6, no. 1 (2014): 217–247.

16. Tatyana Deryugina, "The Fiscal Cost of Hurricanes: Disaster Aid versus Social Insurance," *American Economic Journal: Economic Policy* 9, no. 3 (August 2017): 168–198, https://www.aeaweb.org/articles?id=10.1257/pol.20140296.

17. R. Jisung Park, Joshua Goodman, Michael Hurwitz, and Jonathan Smith, "Heat and Learning," *American Economic Journal: Economic Policy* 12, no. 2 (2020): 306–339.

18. R. Jisung Park, A. Patrick Behrer, and Joshua Goodman, "Learning Is Inhibited by Heat Exposure, Both Internationally and within the United States," *Nature Human Behaviour* 5, no. 1 (May 2020): 19–27, https://doi.org/10.1038/s41562-020-00959-9.

19. Park et al., "Learning Is Inhibited by Heat Exposure."

Chapter Three

1. CA.gov. "Remembering the Camp Fire: The Deadliest and Most Destructive Wildfire in California's History," Accessed August 2023, https://www.fire.ca.gov/our-impact/remembering-the-camp-fire.

2. According to Fortune 500, the top 500 largest companies had a profit of $1.8 trillion in 2022. At a social cost of carbon of approximately $190 per ton, as per the latest Environmental

288   NOTES TO CHAPTER 3

Protection Agency update, and given annual global emissions of 37 billion tons, the net present value of climate damages caused by a given year's emissions can be estimated to be around $7 trillion, the vast majority of which appear to come from hotter temperatures. Nicolas Rapp, "How the Profits of Fortune 500 Companies Stack Up in One Chart," *Fortune*, June 11, 2020, https://fortune.com/longform/fortune-500-companies-profits-compare-2020-chart-data/; US Environmental Protection Agency, "Report on the Social Cost of Greenhouse Gases: Estimates Incorporating Recent Scientific Advance," September, 2022, https://www.epa.gov/system/files/documents/2022-11/epa_scghg_report_draft_0.pdf.

3. Wang Daoping et al., "Economic Footprint of California Wildfires in 2018," *Nature Sustainability* 4, no. 3 (2021): 252–260.

4. John T. Abatzoglou and A. Park Williams, "Impact of Anthropogenic Climate Change on Wildfire across Western US Forests," *Proceedings of the National Academy of Sciences* 113, no. 42 (October 2016): 11770–11775, https://doi.org/10.1073/pnas.1607171113.

5. Marshall Burke et al., "The Changing Risk and Burden of Wildfire in the United States," *Proceedings of the National Academy of Sciences* 118, no. 2 (2021): e2011048118, https://www.pnas.org/doi/10.1073/pnas.2011048118.

6. Katharine Haynes et al., "Wildfires and WUI Fire Fatalities," in *Encyclopedia of Wildfires and WUI Fire Fatalities*, ed. S. Manzello (New York: Springer, 2020), 6.

7. There have been a growing number of large fires in recent years, including the 2009 Victorian bushfires in Australia, with 173 fatalities; the 2017 Portugal forest fires, with 66 fatalities in Pedrógão Grande and 51 fatalities in the Central Region; and the 2018 Mati forest fire in Greece, with 102 fatalities. Burke et al., "Changing Risk and Burden."

8. Marika Katanuma, "New York Has World's Worst Air Pollution as Canada Wildfires Rage," Bloomberg, June 7, 2023, https://www.bloomberg.com/news/articles/2023-06-07/new-york-has-world-s-worst-air-pollution-as-canada-wildfires-rage.

9. Danielle Muoio Dunn, "Canada Wildfires: NY Officials Warn of Health Impacts of Poor Air Quality," Politico, June 7, 2023, https://www.politico.com/news/2023/06/07/wildfire-smoke-new-york-climate-change-00100754#:~:text=NEW%20YORK%20%E2%80%94%20A%20thick%20cloud,pronounced%2C%20city%20officials%20said%20Wednesday.

10. Oliver Milman, "Air Pollution in US from Wildfire Smoke Is Worst in Recent Recorded History," *The Guardian*, June 8, 2023, https://www.theguardian.com/environment/2023/jun/08/air-quality-record-smoke-hazard-wildfire-worst-day-ever-canada-new-york.

11. Xu Yue et al., "Ensemble Projections of Wildfire Activity and Carbonaceous Aerosol Concentrations over the Western United States in the Mid-21st Century," *Atmospheric Environment* 77 (October 2013): 767–780, https://doi.org/10.1016/j.atmosenv.2013.06.003; and *Climate Science Special Report: Fourth National Climate Assessment*, vol. 1, ed. D. J. Wuebbles et al. (Washington, DC: U.S. Global Change Research Program, 2017), https://science2017.globalchange.gov/downloads/CSSR2017_FullReport.pdf.

12. Somini Sengupta, "Why the U.S. Is Backsliding on Clean Air," *New York Times*, June 8, 2023, https://www.nytimes.com/2023/06/08/climate/air-quality-wildfire-smoke.html.

13. Burke et al., "Changing Risk and Burden."

14. T. Deryugina et al., "The Mortality and Medical Costs of Air Pollution: Evidence from Changes in Wind Direction," *American Economic Review* 109, no. 12 (2019): 4178–4219; and Yuyu Chen et al., "Evidence on the Impact of Sustained Exposure to Air Pollution on Life Expectancy

from China's Huai River Policy," *Proceedings of the National Academy of Sciences* 110, no. 32 (2013): 12936–12941.

15. Chen et al. suggested that air pollution may be reducing life expectancy by 0.64 years for every ten microgram per cubic meter increase in PM10. Linear extrapolation should be done with extreme caution; however, to illustrate the magnitude, the implication is that reducing air pollution from an average particulate matter density of 100 micrograms per cubic meter to twenty could increase life expectancy by something on the order of $8 \times 0.64 = 5$ years. In China alone, this would amount to 5×1.4 billion $= 7$ billion life-years. As the authors note, globally, 4.5 billion individuals are exposed to particulate matter concentrations that are at least twice the concentration that the World Health Organization considers safe. Chen et al., "Impact of Sustained Exposure."

16. Burke et al., "Changing Risk and Burden."

17. Of course, many important caveats are in order. It is unclear whether the age-adjusted mortality in either case varies significantly. It could be that direct wildfire deaths include more young people than average mortality due to air pollution, in which case the number of life-years lost may not be as starkly different. As Burke et al. point out, it is also unclear whether the health effects of wildfire smoke-induced PM2.5 varies significantly from the health effects of PM2.5 generally. Finally, as we will discuss in greater detail in the coming chapters, there are many uncertainties regarding the scope for adaptation to wildfire risk—how effective measures such as smoke alerts or air purifiers can be, who can adapt, how quickly, at what cost—all of which may alter the realized impact of wildfire smoke on human health in the future. Burke et al., "Changing Risk and Burden."

18. Burke et al., "Changing Risk and Burden."

19. Tom Chang, Joshua Graff Zivin, Tal Gross, and Matthew Neidell, "Particulate Pollution and the Productivity of Pear Packers," *American Economic Journal: Economic Policy* 8, no. 3 (2016): 141–169.

20. Avraham Ebenstein, Victor Lavy, and Sefi Roth, "The Long-Run Economic Consequences of High-Stakes Examinations: Evidence from Transitory Variation in Pollution," *American Economic Journal: Applied Economics* 8, no. 4 (January 2016): 36–65, https://doi.org/10.1257/app.20150213.

21. Rema Hanna and Paulina Oliva, "The Effect of Pollution on Labor Supply: Evidence from a Natural Experiment in Mexico City," *Journal of Public Economics* 122 (2015): 68–79, https://doi.org/10.1016/j.jpubeco.2014.10.004; and Joshua Graff Zivin and Matthew Neidell, "Temperature and the Allocation of Time: Implications for Climate Change," *Journal of Labor Economics* 32, no. 1 (January 2014): 1–26, https://doi.org/10.1086/671766.

22. Steven Pinker, *The Better Angels of Our Nature: Why Violence Has Declined* (New York: Penguin, 2012).

23. Such spillovers can be geographic, as in the case of wildfire smoke that carries long distances. But they can also operate through hidden economic linkages: across international supply chains or across different sectors of the economy, ranging from manufacturing to construction to retail.

Chapter Four

1. Ken Berger, "Air Conditioning Malfunctions in Steamy Game 1 of NBA Finals," CBS Sports, June 5, 2014, https://www.cbssports.com/nba/news/air-conditioning-malfunctions-in-steamy-game-1-of-nba-finals/.

2. Ben Golliver, "Spurs Apologize for AT&T Center's Malfunctioning Air Conditioning during Game 1," *Sports Illustrated,* June 5, 2014, https://www.si.com/nba/2014/06/06/finals-air-conditioning-att-center-spurs-heat.

3. For instance, see Owen Phillips, "LeBron James Is Indestructible," FiveThirtyEight, ABC News, May 24, 2017, https://fivethirtyeight.com/features/lebron-james-is-indestructible/.

4. Energy Information Administration, *Residential Energy Consumption Survey,* 1984, https://www.eia.gov/consumption/residential/data/archive/pdf/DOE%20EIA-0314(84).pdf.

5. ESPN 30 for 30, *Celtics/Lakers, Best of Enemies,* directed by J. Podhoretz, starring Kareem Abdul-Jabbar, Danny Ainge, Tiny Archibald, and Magic Johnson, June 13, 2017, https://www.imdb.com/title/tt7009262/.

6. Bob Ryan, "In '84, It Was Hot Stuff," *Boston Globe,* June 7, 2009, http://archive.boston.com/sports/basketball/celtics/articles/2009/06/07/in_84_it_was_hot_stuff/.

7. Brett Pollakoff, "With Finals Tied 2–2, a Historical Look at Game 5 Results," NBC Sports, June 16, 2013, https://nba.nbcsports.com/2013/06/16/with-finals-tied-2-2-a-historical-look-at-game-5-results/.

8. ESPN 30 for 30, *Celtics/Lakers, Best of Enemies.*

9. Author's calculations based on data in R. Jisung Park, "Hot Temperature and High-Stakes Performance," *Journal of Human Resources* 57, no. 2 (2022): 400–434, https://doi.org/10.3368/jhr.57.2.0618-9535r3.

10. University of the State of New York, State Education Department, "New York State Regents Exam Schedule, June 2014," https://www.nassauboces.org/cms/lib/NY01928409/Centricity/Domain/37/June%202014%20Updated%20Regents%20Schedule.pdf.

11. Park, "Hot Temperature and High-Stakes Performance."

12. This is because the modal New York City public school student in the data I used is a Hispanic female.

13. A rich economics literature has documented the value of a high school diploma on the job market. While there are important debates about the role of learning and what economists call human capital versus signaling of underlying ability, many studies now suggest that even among those with similar backgrounds, the "sheepskin effect" of simply having the diploma can confer substantial earnings benefits on the order of 10 to 20 percent. See David A. Jaeger and Marianne E. Page, "Degrees Matter: New Evidence on Sheepskin Effects in the Returns to Education," *Review of Economics and Statistics* 78, no. 4 (1996): 733–740.

14. Joshua Goodman, Oded Gurantz, and Jonathan Smith. "Take Two! SAT Retaking and College Enrollment Gaps." *American Economic Journal: Economic Policy* 12, no. 2 (2020): 115–158.

15. Anthony Ramirez, "Reading and Writing, Sweating and Sweltering," *New York Times,* June 15, 2005, https://www.nytimes.com/2005/06/15/nyregion/reading-and-writing-sweating-and-sweltering.html.

16. Lynette Holloway, "A HEAT WAVE SIZZLES: THE SCHOOLS; Without Air, Sweating Starts Early for Students," *New York Times,* July 7, 1999, https://www.nytimes.com/1999/07/07/nyregion/a-heat-wave-sizzles-the-schools-without-air-sweating-starts-early-for-students.html.

17. Natasha Korecki, "That Time Mike Bloomberg Installed an AC Unit in His SUV," Politico, February 21, 2020, https://www.politico.com/news/2020/02/21/mike-bloomberg-income-inequality-116622.

NOTES TO CHAPTER 5   291

18. Ryan, "In '84, It Was Hot Stuff."

19. In response to reporters' questions about his resorting to an oxygen mask, Kareem Abdul-Jabbar retorted, "I suggest that you go to your local steam bath, do 100 pushups with all your clothes on, then try to run back and forth for 48 minutes. The game was in slow motion. It was like we were running in mud." Kareem Abdul-Jabbar, "Kareem Abdul-Jabbar: Miami Fans, Stop Whining about the Fault in Our Air Conditioning," *Time*, June 7, 2014, https://time.com /2843530/kareem-abdul-jabbar-miami-nba-finals/.

Chapter Five

1. Sarah Vander Schaaff, "Lots of Americans Have a Fear of Flying. There Are Ways to Overcome the Anxiety Disorder," *Washington Post*, October 12, 2019, https://www.washingtonpost .com/health/lots-of-americans-have-a-fear-of-flying-there-are-way-to-overcome-the-anxiety -disorder/2019/10/11/d4746d84-d338-11e9-86ac-0f250cc91758_story.html.

2. National Safety Council, "Protect Yourself and Loved Ones by Addressing Roadway Risks," accessed June 2023, https://www.nsc.org/road-safety/safety-topics/fatality-estimates#:~:text =In%202019%2C%20an%20estimated%2038%2C800,2%25%20decrease%20over%202018%20 figures. Over 4.4 million were seriously injured.

3. National Weather Service, "Summary of Natural Hazard Statistics for 2019 in the United States," September 21, 2021, https://www.weather.gov/media/hazstat/sum19.pdf.

4. Centers for Disease Control and Prevention, "Road Traffic Injuries and Deaths—A Global Problem," January 10, 2023, https://www.cdc.gov/injury/features/global-road-safety /index.html.

5. According to a study from the University of California, Davis, each mature lemon tree produces around 900 boxes of lemons per acre (each box is approximately fifty pounds of lemons), and the cost of picking and hauling each fifty-pound box is approximately $2.50/box, which at benefits-inclusive labor costs of $11.70 per worker-hour suggests that a worker is picking roughly 4.6 boxes or 1,200 lemons per hour. Alternatively, twenty worker-hours per acre to pick 900 boxes implies 2,250 lemons per worker per hour. Neil V. O'Connel et al., "Sample Costs to Establish an Orchard and Produce Lemons," University of California Cooperative Extension, 2010, https://coststudyfiles.ucdavis.edu/uploads/cs_public/57/c4/57c4611c-f2bb-4bde-9b77 -a343ca9f0a62/lemonvs10.pdf.

6. United Farm Workers, "Death in the Fields—California Farmworkers Suffer More Heat Deaths and Illnesses than Any Other Workers in Outdoor Industries," November 19, 2015, https://ufw.org/The-Desert-Sun-Death-in-The-Fields-CALIFORNIA-FARMWORKERS -SUFFER-MORE-HEAT-DEATHS-AND-ILLNESSES-THAN-ANY-OTHER-WORKERS -IN-OUTDOOR-INDUSTRIES/. This story is adapted from reporting by the *Desert Sun*.

7. Bureau of Labor Statistics, "National Census of Fatal Occupational Injuries in 2021," December 16, 2022, https://www.bls.gov/news.release/pdf/cfoi.pdf.

8. Bureau of Labor Statistics, "National Census of Fatal Occupational Injuries in 2021."

9. Bureau of Labor Statistics, "43 Work-Related Deaths Due to Environmental Heat Exposure in 2019," *Economics Daily*, September 1, 2022, https://www.bls.gov/opub/ted/2021/43 -work-related-deaths-due-to-environmental-heat-exposure-in-2019.htm.

10. Department for Business, Energy & Industrial Strategy, "Cooling in the UK" (London: The National Archives, August 2021), https://assets.publishing.service.gov.uk/government /uploads/system/uploads/attachment_data/file/1019896/cooling-in-uk.pdf.

11. For data on Indian household air conditioner penetration, see Michael Greenstone, "India's Air-Conditioning and Climate Change Quandary," *New York Times*, October 26, 2016, https://www.nytimes.com/2016/10/27/upshot/indias-air-conditioning-and-climate-change -quandary.html. Urban areas of India appear to have significantly higher rates of air-conditioning, with one survey suggesting 43 percent of households owing air-conditioning units. Nikhil Ghanekar, "Indians Will Buy 300 Mn ACs in 20 Years. And They're Choosing Brands & Features over Energy Efficiency," *India Spend*, May 4, 2021, https://www.indiaspend.com/pollution /indians-will-buy-300-mn-acs-in-20-years-and-theyre-choosing-brands-features-over-energy -efficiency-746363. Ranking the world's hottest cities is not an exact science, given alternative available metrics, but cities such as New Delhi and Kolkata are among the hottest of major population centers. https://en.wikipedia.org/wiki/List_of_cities_by_average_temperature.

12. Nearly 70 percent of these cases were men, and the elderly (those ages 65 and above) account for 36 percent of the total. Centers for Disease Control and Prevention, "Heat-Related Illness" (Fact Sheet), accessed August 7, 2022, https://www.cdc.gov/pictureofamerica/pdfs /Picture_of_America_Heat-Related_Illness.pdf.

13. Centers for Disease Control and Prevention, "Heat-Related Illness."

14. Provisional count of deaths by leading causes of death, United States, 2021 (National Center for Health Statistics, "Table. Provisional Count of Deaths by Leading Causes of Death, United States, 2021," April 12, 2022, https://www.cdc.gov/nchs/data/health_policy/provisional -leading-causes-of-death-for-2021.pdf).

15. Eric Klinenberg, *Heat Wave: A Social Autopsy of Disaster in Chicago* (Chicago: University of Chicago Press, 2015).

16. M. C. Sarofim et al., "Temperature-Related Death and Illness," in *The Impacts of Climate Change on Human Health in the United States: A Scientific Assessment* (Washington, DC: US Global Change Research Program, 2016), 43–68.

17. Olivier Deschenes, "Temperature, Human Health, and Adaptation: A Review of the Empirical Literature," *Energy Economics* 46 (2014): 606–619.

18. Age of infrastructure and general lack of ventilation and air-conditioning could have had something to do with the relatively large impacts of heat on mortality estimated in many Western European settings. The historic European heat wave of 2003 is estimated to have caused anywhere between 20,000 and 40,000 excess deaths across France, Italy, Portugal, Spain, and Germany. That would be at least an order of magnitude more deaths in a single heat wave than all heat-related deaths in the United States each year using the classification method. For reference, the population of France, Italy, Portugal, Spain, and Germany combined is roughly on par with that of the United States. Tom Kosatsky, "The 2003 European Heat Waves," *Eurosurveillance* 10, no. 7 (2005), https://doi.org/10.2807/esm.10.07.00552-en; and Abderrazak Bouchama, "The 2003 European Heat Wave," *Intensive Care Medicine* 30, no. 1–3 (2004), https://doi .org/10.1007/s00134-003-2062-y.

19. Olivier Deschênes and Michael Greenstone, "Climate Change, Mortality, and Adaptation: Evidence from Annual Fluctuations in Weather in the US," *American Economic Journal: Applied Economics* 3, no. 4 (2011): 152–185.

NOTES TO CHAPTER 5   293

20. They look at how variation in temperature within a county affects county-age-specific mortality rates, controlling also for annual variation in other unobserved factors that may affect health at the state level, which may happen to be correlated with temperature, including any long-term trends in both population health and average temperatures.

21. Given the nature of the mortality data, such approximations are not advisable for more than a heuristic indication of magnitude. To translate a change in annual rate to daily rates requires many assumptions, including those regarding stability in daily mortality rates across the year and intertemporal mortality displacement within the year, among other things. It also requires the assumption that such a hot day is replacing a milder, "optimal" temperature day, which may or may not correspond to what typically happens in hotter years, depending on the location.

22. To be precise, these estimates are for replacing those 90°F days with more optimal 60°F days. Of course, a hotter summer doesn't take ideal-temperature days and turn them into heat waves; it likely bumps 85°F days into the 90°F category, 75°F days into the 85°F category, and so on. Moreover, rarely is it the case that the entire US population is hit with the same pulse of hotter-than-average weather. Factor one might mean I am overstating things a bit; factor two, possibly that I am understating things (i.e., How many more people would die if emergency rooms were flooded all at once?). What I am of course not counting are the other less extreme but nevertheless unhealthily hot days, for instance in the mid- to upper 80s (°F).

23. Garth Heutel, Nolan H. Miller, and David Molitor, "Adaptation and the Mortality Effects of Temperature across US Climate Regions," *Review of Economics and Statistics* 103, no. 4 (2021): 740–753, https://doi.org/10.1162/rest_a_00936.

24. Given an elderly population of 54 million at the time of their study, the mortality rates mentioned previously, and the prevalence of days in the 80s and lower 90s (°F) in the United States during the time of their study, as well as prevailing patterns of adaptation to heat, I estimate that roughly 7,500 elderly deaths may have been attributable to hotter temperature in the United States. This may be an overestimate if we believe the US economy has become better adapted to heat over time. Conversely, it may be an underestimate given the increase in hotter temperatures relative to the study period (1992–2013).

25. This study, as well as the studies by Deschênes and by Greenstone and Heutel et al., use daily average temperature. Olivier Deschênes, "Temperature Variability and Mortality: Evidence from 16 Asian Countries," *Asian Development Review* 35, no. 2 (October 2018): 1–30, https://doi.org/10.1162/adev_a_00112.

26. Tamma Carleton et al. "Valuing the Global Mortality Consequences of Climate Change Accounting for Adaptation Costs and Benefits," *Quarterly Journal of Economics* 137, no. 4 (2022): 2037–2105.

27. An emphasis must be placed on the phrase "back-of-the-envelope." I arrive at this approximation by taking Carleton et al.'s estimate of the marginal effect of a day above 30°C on elderly (age 65 and above) mortality rates, which is reported as being +2.77 per 100,000 people, multiplied by the average number of days in that temperature range for their study sample (32.6 days), times the number of people in that age bracket in the global population (8.1 billion). This leads to an estimate of 707,000.

28. Qi Zhao et al., "Global, Regional, and National Burden of Mortality Associated with Non-optimal Ambient Temperatures from 2000 to 2019: A Three-Stage Modelling Study," *The Lancet Planetary Health* 5, no. 7 (2021): e415–e425.

294 NOTES TO CHAPTER 6

29. Centers for Disease Control and Prevention, "HIV Surveillance Report: Diagnosis of HIV Infection in the United States and Dependent Areas," vol. 33, May 2020, https://www.cdc .gov/hiv/statistics/overview/index.html#:~:text=Worldwide%2C%20there%20were%20 about%201.5,the%20start%20of%20the%20epidemic; Centers for Disease Control and Prevention, "Malaria's Impact Worldwide," December 16, 2021, https://www.cdc.gov/malaria/malaria _worldwide/impact.html#:~:text=In%202020%2C%20malaria%20caused%20an,clinical%20 episodes%2C%20and%20627%2C000%20deaths.

30. This estimate is derived from survey data that measures core vitals, including blood sugar levels. Specifically, diagnosed diabetes cases were based on self-report, and undiagnosed diabetes cases were based on fasting plasma glucose and A1C levels among people self-reporting no diabetes. Centers for Disease Control and Prevention "Prevalence of Both Diagnosed and Undiagnosed Diabetes," September 30, 2022, https://www.cdc.gov/diabetes/data/statistics-report /diagnosed-undiagnosed-diabetes.html.

31. Our data comes from internal documents shared with us by Cal/OSHA staff. We are grateful for their cooperation in this work.

32. Carleton et al., "Valuing the Global Mortality Consequences."

33. William Booth, "Cerberus. Charon. Gary? Heat Wave Naming Debate Intensifies," *Washington Post*, July 20, 2023, https://www.washingtonpost.com/world/2023/07/20/europe-heat -wave-name-cerebus-charon/.

Chapter Six

1. According to the International Labour Organization (ILO), agriculture alone employs 1.3 billion people. ILO, "Agriculture: A Hazardous Work," accessed May 2020, https://www.ilo.org /safework/areasofwork/hazardous-work/WCMS_110188/lang--en/index.htm.

2. Nicole Maestas et al., "The Value of Working Conditions in the United States and Implications for the Structure of Wages" (working paper 25204, National Bureau of Economic Research, 2018), https://doi.org/10.3386/w25204.

3. R. Jisung Park et al., "Heat and Learning," *American Economic Journal: Economic Policy* 12, no. 2 (2020): 306–339.

4. Daron Acemoglu and Melissa Dell, "Productivity Differences between and within Countries," *American Economic Journal: Macroeconomics* 2, no. 1 (January 2010): 169–188, https://doi .org/10.1257/mac.2.1.169; and William D. Nordhaus, "Revisiting the Social Cost of Carbon," *Proceedings of the National Academy of Sciences* 114, no. 7 (2017): 1518–1523, https://doi.org/10 .1073/pnas.1609244114.

5. Melissa Dell, Benjamin F. Jones, and Benjamin A. Olken, "What Do We Learn from the Weather? The New Climate-Economy Literature," *Journal of Economic Literature* 52, no. 3 (January 2014): 740–798, https://doi.org/10.1257/jel.52.3.740; and Melissa Dell, Benjamin F Jones, and Benjamin A Olken, "Temperature Shocks and Economic Growth: Evidence from the Last Half Century," *American Economic Journal: Macroeconomics* 4, no. 3 (January 2012): 66–95, https://doi.org/10.1257/mac.4.3.66.

6. Samuel Huntington, *Civilization and Climate* (New Haven, CT: Yale University Press, 1915).

7. For instance, this was not a double-blind study. For many of these subjects, this was simply an extension of a battery of tests and experiments that they had been taking part in, all of which

were aimed at testing the limits of human performance under high temperature. So they were likely well aware of what the purpose of the experiment was, meaning that the Hawthorne effect (i.e., effects that arise due to subjects' awareness of being monitored) and other forms of framing bias may have affected the findings.

8. Olli Seppänen, William J. Fisk, and Q. H. Lei, "Effect of Temperature on Task Performance in Office Environment" (paper, 5th International Conference on Cold Climate Heating, Ventilating and Air Conditioning, Moscow, May 21–24, 2006); and Geoffrey Heal and Jisung Park, "Reflections—Temperature Stress and the Direct Impact of Climate Change: A Review of an Emerging Literature," *Review of Environmental Economics and Policy* 10, no. 2 (2016): 347–362.

9. Seppänen et al., "Effect of Temperature."

10. Gavin C. Donaldson, William R. Keatinge, and Richard D. Saunders, "Cardiovascular Responses to Heat Stress and Their Adverse Consequences in Healthy and Vulnerable Human Populations," *International Journal of Hyperthermia* 19, no. 3 (2003): 225–235, https://doi.org/10.1080/0265673021000058357.

11. Marcus E. Raichle and Mark A. Mintun, "Brain Work and Brain Imaging," *Annual Review of Neuroscience* 29 (2006): 449–476, https://doi.org/10.1146/annurev.neuro.29.051605.112819.

12. Lars Nybo and Niels H. Secher, "Cerebral Perturbations Provoked by Prolonged Exercise," *Progress in Neurobiology* 72, no. 4 (2004): 223–261.

13. Centers for Disease Control and Prevention, "The Syphilis Study at Tuskegee Timeline," December 5, 2022, https://www.cdc.gov/tuskegee/timeline.htm.

14. I will not encumber the casual reader with the particulars here, and instead direct them toward accessible guides to this evolving econometric landscape, including Joshua D. Angrist and Jörn-Steffen Pishke's *Mastering Metrics: The Path from Cause to Effect* (2015) and Scott Cunningham's *Causal Inference: The Mixtape* (2021). While discussion of the specific conditions that need to be met in order to claim causality and to be confident in the statistical significance of any given relationship documented is beyond the scope of this book, it is worth emphasizing the importance of checking these conditions diligently, especially when working with large data sets that can, without such diligence, easily give off an impression of false statistical significance, and therefore lead to misplaced confidence.

15. In the paper "Hot Temperature and High-Stakes Performance," which was a chapter of my PhD dissertation, I also control for possible differences in average difficulty across subjects, as well as the possible correlation between timing of an exam later in the day or later in the week with fatigue that might be spuriously correlated with hotter temperature. R. Jisung Park, "Hot Temperature and High-Stakes Performance," *Journal of Human Resources* 57, no. 2 (2022): 400–434.

16. Paul Froom et al., "Heat Stress and Helicopter Pilot Errors," *Journal of Occupational Medicine* 35, no. 7 (1993): 720–724.

17. Joshua Graff Zivin, Yingquan Song, Qu Tang, and Peng Zhang, "Temperature and High-Stakes Cognitive Performance: Evidence from the National College Entrance Examination in China," *Journal of Environmental Economics and Management* 104 (2020): 102365.

18. Ruixue Jia and Hongbin Li, "Access to Elite Education, Wage Premium, and Social Mobility: Evidence from China's College Entrance Exam" (working paper no. 577, Stanford Center for International Development, 2016), https://kingcenter.stanford.edu/sites/g/files/sbiybj16611/files/media/file/577wp_1_0.pdf. The extent to which financial returns to educational

attainment are a function of true human capital accumulation or other factors, like signaling of underlying ability, remains debated. So one should extrapolate the income differences with caution.

19. Marshall Burke et al., "Game, Sweat, Match: Temperature and Elite Worker Productivity" (working paper no. 31650, National Bureau of Economic Research, 2023), https://www.nber.org/papers/w31650#:~:text=The%20effect%20of%20hot%20temperatures,between%20temperature%20and%20economic%20output.

20. There are of course limits to extrapolation. The most immediate implications are for other sports to take place outdoors: soccer, football, cricket, and golf. Indeed, new data from Chinese archery competitions echoes these findings. For instance, see Yun Qiu and Jinhua Zhao, "Adaptation and the Distributional Effects of Heat: Evidence from Professional Archery Competitions," *Southern Economic Journal* 88, no. 3 (2022): 1149–1177, https://doi.org/10.1002/soej.12553. However, this also highlights the importance of thinking about context. To some extent, professional sports is inherently a zero-sum affair. So as long as temperature is affecting each competitor's abilities equally (and there are no serious health consequences), we may not care so much. But not all things in life are zero-sum. In fact, the majority of productive economic activity is decidedly not zero-sum. Tennis balls are known to be affected by hotter temperatures, which makes them bouncier. This can affect players' strategy. Bouncier balls mean a faster pace of play and may favor players with strong forehands who can make the most of the quickened pace. It's not clear whether this mutes the effect for some players and not others. Perhaps, more importantly, there is still very limited scope for adaptation, as with the New York City study. Players can't request an air-conditioned court or postponement on account of the temperature. Aleks Szymanski, "How the Weather Impacts a Tennis Match," Tennisletics, November 26, 2018, https://www.tennisletics.com/blog/how-the-weather-impacts-a-tennis-match/; Andrew Beaton, "Summer Heat Gives Tennis Balls an Extra Bounce," *Wall Street Journal*, September 2, 2015, https://www.wsj.com/articles/heat-gives-tennis-balls-an-extra-bounce-1441209302.

21. Gerard P. Cachon, Santiago Gallino, and Marcelo Olivares. "Severe Weather and Automobile Assembly Productivity" (Columbia Business School Research Paper 12/37, 2012).

22. Even if plants did manage to make up for lost production by working overtime, this would entail additional costs, and the volatility of production itself would impose a net cost on the company over time.

23. Philippe Kabore and Nicholas Rivers, "Manufacturing Output and Extreme Temperature: Evidence from Canada" (working paper, 20-11, Clean Economy Working Paper Series, 2020), https://institute.smartprosperity.ca/sites/default/files/WP_manufacturing%20Output.pdf.

24. This requires assumptions similar to the heat-mortality extrapolation from chapter 5. With those caveats, I calculate that, with 260 workdays out of the year, linear damages, and no intertemporal substitution of production across weeks within the year, an annualized impact of -0.11 percent translates to $-0.11 \times 260 = -28.6$ percent of daily production.

25. While cooling large manufacturing plants may impose nontrivial economic costs, it's not as if the technologies to do so are not widely available.

26. Eswaran Somanathan et al., "The Impact of Temperature on Productivity and Labor Supply: Evidence from Indian Manufacturing," *Journal of Political Economy* 129, no. 6 (2021): 1797–1827.

NOTES TO CHAPTER 7   297

27. Peng Zhang et al., "Temperature Effects on Productivity and Factor Reallocation: Evidence from a Half Million Chinese Manufacturing Plants," *Journal of Environmental Economics and Management* 88 (2018): 1–17, https://doi.org/10.1016/j.jeem.2017.11.001.

28. Jawad M. Addoum, David T. Ng, and Ariel Ortiz-Bobea. "Temperature Shocks and Industry Earnings News," *Journal of Financial Economics*, forthcoming, http://dx.doi.org/10.2139/ssrn.3480695.

29. See the influential work by Acemoglu, Johnson, and Robinson, as well as Rodrik, Subramanian, and Trebbi, on the role of institutions versus geography in influencing long-term economic growth: Daron Acemoglu, Simon Johnson, and James A. Robinson, "The Colonial Origins of Comparative Development: An Empirical Investigation," *American Economic Review* 91, no. 5 (2001): 1369–1401; Dani Rodrik, Arvind Subramanian, and Francesco Trebbi, "Institutions Rule: The Primacy of Institutions over Geography and Integration in Economic Development," *Journal of Economic Growth* 9 (2004): 131–165.

30. Solomon M. Hsiang, "Temperatures and Cyclones Strongly Associated with Economic Production in the Caribbean and Central America," *Proceedings of the National Academy of Sciences* 107, no. 35 (2010): 15367–15372.

31. Dell, Jones, and Olken, "Temperature Shocks and Economic Growth."

32. Eric Fesselmeyer, "The Impact of Temperature on Labor Quality: Umpire Accuracy in Major League Baseball," *Southern Economic Journal* 88, no. 2 (2021): 545–567; Steffen Künn, Juan Palacios, and Nico Pestel, "The Impact of Indoor Climate on Human Cognition: Evidence from Chess Tournaments" (working paper, 2019), https://conference.iza.org/conference_files/environ_2019/palacios_j24419.pdf.

33. For instance, as DJO put it: "Our estimates show large, negative effects of higher temperatures on growth, but only in poor countries.... In rich countries, changes in temperature do not have a robust, discernable effect on growth." Dell, Jones, and Olken, "Temperature Shocks and Economic Growth."

34. Solomon Hsiang et al., "Estimating Economic Damage from Climate Change in the United States," *Science* 356, no. 6345 (2017): 1362–1369.

35. Will Daniel, "Apple Was the Most Profitable Company on the Fortune 500 List This Year. These Are the Biggest Profit Generators, and What That Means About American Business," *Fortune*, May 24, 2022, https://fortune.com/2022/05/24/fortune-500-most-profitable-companies-apple-berkshire-amazon-google-microsoft/.

Chapter Seven

1. Caitlin Dickerson, "Inside the Refugee Camp on America's Doorstep," *New York Times*, October 23, 2020, https://www.nytimes.com/2020/10/23/us/mexico-migrant-camp-asylum.html.

2. Under US law, a "refugee" is a person who is unable or unwilling to return to his or her home country because of a "well-founded fear of persecution" due to race, membership in a particular social group, political opinion, religion, or national origin. This definition is based on the United Nations 1951 Convention and 1967 Protocols relating to the Status of Refugees, which the United States became a party to in 1968. Following the Vietnam War and the US

experience resettling Indochinese refugees, Congress passed the Refugee Act of 1980, which incorporated the Convention's definition into US law and provides the legal basis for today's US Refugee Admissions Program.

3. Carrie Kahn, "'Remain in Mexico' Policy's End Brings Renewed Hope to Asylum Seekers," NPR, February 14, 2021, https://www.npr.org/2021/02/14/967807189/remain-in-mexico -policys-end-brings-renewed-hope-to-asylum-seekers.

4. Gaia Vince, "The Century of Climate Migration: Why We Need to Plan for the Great Upheaval," *The Guardian*, August 18, 2022, https://amp.theguardian.com/news/2022/aug/18 /century-climate-crisis-migration-why-we-need-plan-great-upheaval; Jessie Leung, "Climate Crisis Could Displace 1.2 Billion People by 2050, Report Warns," CNN, September 10, 2020, https://www.cnn.com/2020/09/10/world/climate-global-displacement-report-intl-hnk-scli -scn/index.html.

5. UN Environment Programme, "UN Environment and European Union Call for Stronger Action against Climate-Change-Related Security Threat," UNEP.org, June 22, 2018, https:// www.unep.org/news-and-stories/story/un-environment-and-european-union-call-stronger -action-against-climate.

6. Marshall Burke, Solomon M. Hsiang, and Edward Miguel, "Climate and Conflict," *Annual Review of Economics* 7, no. 1 (2015): 577–617; and Robert McLeman, "International Migration and Climate Adaptation in an Era of Hardening Borders," *Nature Climate Change* 9, no. 12 (2019): 911–918.

7. Studies have shown that even the perceived threat of violence can be psychologically harmful to children and adults. For instance, Patrick Sharkey, Amy Ellen Schwartz, Ingrid Gould Ellen, and Johanna Lacoe find that New York City public school students who were exposed to violent crime on their own block during the week leading up to an exam performed significantly worse on standardized exams, even though they were not themselves victims of physical violence. Patrick Sharkey et al., "High Stakes in the Classroom, High Stakes on the Street: The Effects of Community Violence on Students' Standardized Test Performance," *Sociological Science* 1 (2014): 199.

8. Urban Institute, "Criminal Justice Expenditures: Police, Corrections, and Courts," accessed September 2022, https://www.urban.org/policy-centers/cross-center-initiatives/state -and-local-finance-initiative/state-and-local-backgrounders/criminal-justice-police-corrections -courts-expenditures.

9. Eurostat, "Member States' Spending on Public Order and Safety," August 16, 2017, https:/ /ec.europa.eu/eurostat/web/products-eurostat-news/-/DDN-20170816-1. It may be worth noting that this level of expenditure is roughly equivalent to the costs associated with a €100 to €150 per ton carbon tax, given 2022 levels of emissions per capita.

10. World Bank, "Intentional Homicides per 100,000 People," 2021, https://data.worldbank .org/indicator/VC.IHR.PSRC.P5?end=2021&locations=ZA&start=2019.

11. It is important to note that the US average masks enormous heterogeneity. For many minority communities, especially in high poverty urban neighborhoods, local homicide rates can be many times the nationwide average and on par with places like El Salvador.

12. The rates of homicide are 1.2, 0.9, 0.6, and 0.3 per 100,000 for the United Kingdom, Germany, South Korea, and Japan, respectively.

NOTES TO CHAPTER 7   299

13. Samuel Huntington, "The Clash of Civilizations," *Foreign Affairs* 72, no. 3 (1993): 22–49.

14. Matthew Ranson, "Crime, Weather, and Climate Change," *Journal of Environmental Economics and Management* 67, no. 3 (2014): 274–302.

15. As with other studies in the new climate-economics literature, the researcher is careful to account not only for potential spurious correlation between geographic factors and temperature—how the US South tends to be both warmer and poorer on average—as well as for seasonality. The latter is an important control given clear seasonality in both rates of crime as well as other factors—like the number of people on vacation or temporarily unemployed—which might change with the seasons and may thus be correlated with temperature. In fact, it is more than that, as the study controls for county by year and state by month fixed effects, which means that these estimates are controlling for not only county-specific average variation in crime rates, but variation in seasonality in crime rates by state.

16. Douglas T. Kenrick and Steven W. MacFarlane, "Ambient Temperature and Horn Honking: A Field Study of the Heat/Aggression Relationship," *Environment and Behavior* 18, no. 2 (1986): 179–191.

17. The authors note that "several of our subjects accompanied their horn honking with other verbal and nonverbal signals of hostility," and that "the highest levels of horn honking were obtained from young male subjects accompanied by groups of other males." Because of the lack of systematic data collection and the relatively small sample sizes, it is difficult to draw clear inferences from these suggestive indicators.

18. Larry Printz, "This Is How Air Conditioning Became Cool," Hagerty Media, July 25, 2016, https://www.hagerty.com/media/maintenance-and-tech/air-conditioning/.

19. Alan S. Reifman, Richard P. Larrick, and Steven Fein, "Temper and Temperature on the Diamond: The Heat-Aggression Relationship in Major League Baseball," *Personality and Social Psychology Bulletin* 17, no. 5 (1991): 580–585.

20. Curtis Craig, Randy W. Overbeek, Miles V. Condon, and Shannon B. Rinaldo, "A Relationship between Temperature and Aggression in NFL Football Penalties," *Journal of Sport and Health Science* 5, no. 2 (2016): 205–210.

21. Peter Baylis, "Temperature and Temperament: Evidence from Twitter," *Journal of Public Economics* 184 (2020): 104161.

22. The researcher uses measures of expressed sentiment derived from prior work, including the AFINN score, as well as Hedonometer and others. Finn Årup Nielsen, "AFINN Project" (GitHub repository, DTU Compute Technical University of Denmark, 2017); and Morgan R. Frank et al., "Happiness and the Patterns of Life: A Study of Geolocated Tweets," *Scientific Reports* 3, no. 1 (2013): 2625.

23. Baylis's findings appear to be consistent with the well-documented relationship between hotter temperature and violent crime. But it is worth noting that very cold days with temperatures below 3°C also appear to reduce sentiment by a similar magnitude: roughly 15 percent of a standard deviation. The fact that Baylis found that colder temperatures induce more profane text than moderate temperatures is intriguing because most work on temperature in crime finds that colder temperatures do not significantly increase criminal activity. One possibility is that while aggressive impulses increase during more extreme temperatures (both hot and cold),

300 NOTES TO CHAPTER 7

cooler temperatures act to limit the opportunities of individuals to interact with each other or to act on that kind of aggression in ways that show up in the data.

24. Mental Health America, "Mental Health in America—Adult Data 2018," accessed March 2022, https://www.mhanational.org/issues/mental-health-america-adult-data-2018.

25. Editorial staff, "Mental Health Matters," *The Lancet Global Health* 8, no. 11 (2020): e1352.

26. Jamie T. Mullins and Corey White, "Temperature and Mental Health: Evidence from the Spectrum of Mental Health Outcomes," *Journal of Health Economics* 68 (2019): 102240.

27. It is worth noting that, in Mullins and White's analysis, they find evidence consistent with disrupted sleep being an important mechanism. Previous literature finds a strong link between sleep quality and quantity and mental illness and suicide. Mullins and White use American Time Use Survey data to show that hotter temperature affects sleep quality and quantity, and that higher daily minimum temperatures appear to be more negatively impactful on their measures of mental health than higher daily maximum temperature, which would be consistent with sleep as an important mechanism.

28. Tamma A. Carleton, "Crop-Damaging Temperatures Increase Suicide Rates in India," *Proceedings of the National Academy of Sciences* 114, no. 33 (2017): 8746–8751.

29. For a review of this fascinating literature, see Burke et al., "Climate and Conflict." For a critique of some of the earlier work on this topic, see H. Buhaug et al., "One Effect to Rule Them All? A Comment on Climate and Conflict," *Climatic Change* 127 (2014): 391–397.

30. Shuaizhang Feng, Alan B. Krueger, and Michael Oppenheimer ("Linkages among Climate Change, Crop Yields and Mexico–US Cross-Border Migration," *Proceedings of the National Academy of Sciences* 107, no. 32 [2010]: 14257–14262) present evidence on the Mexico–US migration due to weather shocks. Anouch Missirian and Wolfram Schlenker ("Asylum Applications Respond to Temperature Fluctuations," *Science* 358, no. 6370 [2017]: 1610–1614) show how European Union asylum applications respond to weather shocks in sending countries.

31. Department of Defense, "Department of Defense Climate Risk Analysis," October 2021, https://media.defense.gov/2021/Oct/21/2002877353/-1/-1/0/DOD-CLIMATE-RISK-ANALYSIS-FINAL.PDF.

32. For a more in-depth discussion of climate migration and where the literature on this topic stands, see Cristina Cattaneo et al., "Human Migration in the Era of Climate Change," *Review of Environmental Economics and Policy* 13, no. 2 (2019): 189–206.

33. A long literature on human migration distinguishes between "survival migration," which is often the last resort of low-wealth families facing adverse shocks, and typically involve temporary relocations over short distances, and "profitable investment migration," whereby (often richer or less destitute) families engage in rural-urban or international migration in search of better economic opportunities, often lasting longer. By one estimate, such long-distance "investment migrations" are about four times as costly as the shorter-distance "survival" variety.

34. Barbora Šedová, Lucia Čizmaziová, and Athene Cook, "A Meta-Analysis of Climate Migration Literature" (Discussion paper no. 29, Center for Economic Policy Analysis, 2021), https://publishup.uni-potsdam.de/opus4-ubp/frontdoor/deliver/index/docId/49982/file/cepa29.pdf.

35. "How Climate Change Can Fuel Wars," *The Economist*, May 23, 2019, https://www.economist.com/international/2019/05/23/how-climate-change-can-fuel-wars.

NOTES TO CHAPTER 8 301

36. I am indebted to the research and clear exposition of these and other relevant legal information of Anthony Heyes and Soodeh Saberian, especially their article "Temperature and Decisions: Evidence from 207,000 Court Cases," *American Economic Journal: Applied Economics* 11, no. 2 (2017): 238–265.

37. A. Patrick Behrer and Valentin Bolotnyy, "Heat, Crime, and Punishment" (Economics working paper 21114, Economic Policy Working Group, 2022), https://www.hoover.org/research/heat-crime-and-punishment.

38. Terry-Ann Craigie, Vis Taraz, and Mariyana Zapryanova, "Temperature and Convictions: Evidence from India" (working paper, Economics: Faculty Publications, Smith College, Northampton, MA, 2023), https://scholarworks.smith.edu/eco_facpubs/105/.

39. Daniel L. Chen and Holger Spamann, "This Morning's Breakfast, Last Night's Game: Detecting Extraneous Influences on Judging" (IAST working paper, 2016); and Ozkan Eren and Naci Mocan, "Emotional Judges and Unlucky Juveniles," *American Economic Journal: Applied Economics* 10, no. 3 (July 2018): 171–205.

40. Anita Mukherjee and Nicholas J. Sanders, "The Causal Effect of Heat on Violence: Social Implications of Unmitigated Heat among the Incarcerated" (working paper no. 28987, National Bureau of Economic Research, 2021), https://www.nber.org/system/files/working_papers/w28987/w28987.pdf.

Part III

1. Lee Kwan Yew, "The East Asian Way—with Air Conditioning," *New Perspectives Quarterly* 26, no. 4 (2009): 111–120.

2. Lucas W. Davis and Paul J. Gertler, "Contribution of Air Conditioning Adoption to Future Energy Use Under Global Warming," *Proceedings of the National Academy of Sciences* 112, no. 19 (2015): 5962–5967.

3. World Bank, "CO_2 Emissions (Metric Tons per Capita), United Kingdom" (Washington, DC: World Resources Institute, 2023), https://data.worldbank.org/indicator/EN.ATM.CO2E.PC.

Chapter Eight

1. South Korea's first health insurance law wasn't passed until 1963, which made it possible for employers to begin providing voluntary health insurance to employees. Columbia University, "South Korea: Summary," accessed September 2022, https://www.publichealth.columbia.edu/research/comparative-health-policy-library/south-korea-summary#:~:text=Korea%20achieved%20universal%20health%20coverage,societies%20into%20a%20single%20insurer.

2. The average speed on the Amtrak Acela is 68 miles per hour (mph). The average speed on the KTX is 180 mph. Neither is quite as fast as France's TGV (198 mph), or Shanhai's Transrapid (267 mph). Maglev.net, "World's Fastest High-Speed Trains in Commercial Operation in 2021," February 5, 2020, https://www.maglev.net/worlds-fastest-high-speed-trains-in-commercial-operation; and Kate Taylor, "I Rode Seoul's Famous Subway System for a Week to See If It's Really the Best in the Word, and Saw Why New York Will Never Catch Up," Insider, February 7,

2020, https://www.businessinsider.com/seouls-subway-system-better-than-new-york-subway-2020-2#however-following-the-signs-it-was-easy-to-get-to-gyeongbokgung-palace-21.

3. According to World Bank data, Ghana's per capita income in 1960 was approximately $1,150 and South Korea's was $1,027. World Bank, "Constant GDP per Capita for Ghana," Federal Reserve Economic Data, September 6, 2023, https://fred.stlouisfed.org/series/NYGDPPCAPKDGHA; World Bank, "Constant GDP per Capita for the Republic of Korea," Federal Reserve Economic Data, September 6, 2023, https://fred.stlouisfed.org/series/NYGDPPCAPKDKOR.

4. Roger Fouquet and Stephen Broadberry, "Seven Centuries of European Economic Growth and Decline," *Journal of Economic Perspectives* 29, no. 4 (2015): 227–244.

5. The thirty-fold comparison uses non-PPP adjusted values from the World Bank. Just between when my sister and I lived there in the late 1990s and now, per capita incomes have nearly tripled. World Bank, "GDP per Capita, PPP (Current International $)—Korea Rep," September 6, 2023, https://data.worldbank.org/indicator/NY.GDP.PCAP.PP.CD?locations=KR.

6. C. An and Barry Bosworth, *Income Inequality in Korea: An Analysis of Trends, Causes and Answers*, Harvard East Asian Monographs, vol. 354 (Cambridge, MA: Harvard University Asia Center, 2013).

7. Paul Johnson and Chris Papageorgiou, "What Remains of Cross-Country Convergence?," *Journal of Economic Literature* 58, no. 1 (2020): 129–175, https://www.aeaweb.org/articles?id=10.1257/jel.20181207.

8. J. Bradford DeLong, "Growth in the World Economy, ca. 1870–1990," in *Economic Growth in the World Economy* (Tübingen: Mohr/Siebeck, 1992). While accurate statistics are harder to compile the further back in history we go, such growth rates were likely faster still than the centuries prior.

9. Max Roser, Esteban Ortiz-Ospina, and Hannah Ritchie, "Life Expectancy," Our World in Data, October 2019, https://ourworldindata.org/life-expectancy. Life expectancy in Germany increased from 66 years in 1950 to 81 years in 2015.

10. For a nuanced discussion of the decline of violence over the span of recorded human history, see Stephen Pinker, *The Better Angels of Our Nature: Why Violence Has Declined* (New York: Penguin, 2011).

11. J. Bradford DeLong, *Slouching towards Utopia: An Economic History of the Twentieth Century* (New York: Basic Books, 2022).

12. Johnson and Papageorgiou, "What Remains of Cross-Country Convergence?"

13. To put this in perspective, the average American in 1900 is estimated to have had a material standard of living of around $12,400 in today's dollars. Angus Maddison, in *The World Economy* (Paris: Development Centre of the Organisation for Economic Co-operation and Development, 2006), calculates US per capita GDP to have been $5,300 in 1990 dollars which, adjusting for inflation using the Bureau of Labor Statistics Consumer Price Index deflator, comes to $12,400.

14. While extreme wealth concentration at the top (the top 1 percent, or even the top 0.1 percent, for instance) receives much popular attention, it is important to also note the rise in inequality among the rest of the population—the other 99 percent—for whom wage earnings remain the primary source of income.

NOTES TO CHAPTER 8 303

15. David H. Autor, "Skills, Education, and the Rise of Earnings Inequality among the 'Other 99 Percent,'" *Science* 344, no. 6186 (2014): 843–851.

16. There are some notable exceptions to these trends, including in some Latin American countries, like Venezuela, where income inequality has declined, albeit from a very high starting point. Many African countries have had very high levels of economic inequality throughout this period.

17. OECD, "In It Together: Why Less Inequality Benefits All," figure 1, May 21, 2015, https://doi.org/10.1787/9789264235120-en.

18. World Inequality Lab, Global Inequality Data: 2020 update, accessed July 2023, https://wid.world/news-article/2020-regional-updates/.

19. Daron Acemoglu and Pascual Restrepo, "Tasks, Automation, and the Rise in US Wage Inequality," *Econometrica* 90, no. 5 (2022): 1973–2016; David Autor, "The Work of the Future: Shaping Technology and Institutions" (lecture, UBS Center, December 2, 2019), https://www.youtube.com/watch?v=0FbFjCmZQRM&list=PLi_W2VBEkN127AIOJkMCJLgasCHjBcQWe&index=5.

20. Raj Chetty et al. "Where Is the Land of Opportunity? The Geography of Intergenerational Mobility in the United States," *Quarterly Journal of Economics* 129, no. 4 (November 2014): 1553–1623, https://doi.org/10.1093/qje/qju022.

21. Raj Chetty et al., "The Association between Income and Life Expectancy in the United States, 2001–2014," *JAMA* 315, no. 16 (2016): 1750–1766.

22. Average life expectancy for Americans (both sexes) is 77 years in 2023 according to the World Bank. For Kenyans, average life expectancy is 63 years. World Population Review, "Life Expectancy by Country 2023," accessed March 2023, https://worldpopulationreview.com/country-rankings/life-expectancy-by-country.

23. Hoops Geek, "The Average Height of NBA Players from 1952–2022," July 4, 2022, https://thehoopsgeek.com/average-nba-height/.

24. National Weather Service, "Average Monthly & Annual Temperatures at Central Park," May 9, 2021, https://www.weather.gov/media/okx/Climate/CentralPark/monthlyannualtemp.pdf; Weather Spark, "Climate and Average Weather Year Round in Bangkok, Thailand," accessed March 2023, https://weatherspark.com/y/113416/Average-Weather-in-Bangkok-Thailand-Year-Round.

25. Garth Heutel, Nolan H. Miller, and David Molitor, "Adaptation and the Mortality Effects of Temperature across US Climate Regions," *Review of Economics and Statistics* 103, no. 4 (2021): 740–753, https://doi.org/10.1162/rest_a_00936.

26. Melissa Dell, Benjamin F. Jones, and Benjamin A. Olken, "Temperature and Income: Reconciling New Cross-Sectional and Panel Estimates," *American Economic Review* 99, no. 2 (2009): 198–204, https://pubs.aeaweb.org/doi/pdf/10.1257/aer.99.2.198.

27. The magnitude is nontrivial. Even allowing for households to sort according to heterogeneous preferences, they are willing to pay over 1 percent of total income to avoid moderate warming by 2050. One percent may not sound like a lot, but that is over $70 billion in the United States in 2020 dollars. David Albouy, Walter Graf, Ryan Kellogg, and Hendrik Wolff, "Climate Amenities, Climate Change, and American Quality of Life," *Journal of the Association of Environmental and Resource Economists* 3, no. 1 (2016): 205–246; Paramita Sinha, Martha L. Caulkins,

and Maureen L. Cropper, "Household Location Decisions and the Value of Climate Amenities," *Journal of Environmental Economics and Management* 92 (2018): 608–637.

28. David Maddison and Andrea Bigano, "The Amenity Value of the Italian Climate," *Journal of Environmental Economics and Management* 45, no. 2 (2003): 319–332; Katrin Rehdanz and David Maddison, "Climate and Happiness," *Ecological Economics* 52, no. 1 (2005): 111–125.

29. Jisung Park et al., "Households and Heat Stress: Estimating the Distributional Consequences of Climate Change," *Environment and Development Economics* 23, no. 3 (2018): 349–368, https://doi.org/10.1017/S1355770X1800013X.

30. Brad Plumer and Nadja Popovich, "How Decades of Racist Housing Policy Left Neighborhoods Sweltering," *New York Times*, August 24, 2020, https://www.nytimes.com/interactive/2020/08/24/climate/racism-redlining-cities-global-warming.html.

31. Plumer and Popovich, "Decades of Racist Housing Policy."

Chapter Nine

1. DQYDJ, "Average, Median, Top 1%, and all United States Salary Percentiles," September 30, 2021, https://dqydj.com/2020-average-median-top-salary-percentiles/. The data on this site was taken from the 2020 Census Integrated Public Use Microdata Series.

2. For a great analysis of the growing importance of social skills in the labor market, see David Deming's work, including "The Growing Importance of Social Skills in the Labor Market," *Quarterly Journal of Economics* 132, no. 4 (November 2017): 1593–1640, https://doi.org/10.1093/qje/qjx022.

3. It is worth noting that, not only is it the case that average earnings between those with and without college degrees is growing, it is also the case that the divergence in wages (for a given occupation) between high-paying firms and low-paying firms seems to be growing, and the importance of education in finding and securing these high-paying firm positions seems to be growing as well. So my decision to look only within Amazon, which was motivated primarily to illustrate stark juxtapositions with the same company, may in some sense undersell the true extent of inequality because business development representatives, warehouse workers, and janitors at smaller, less profitable firms may experience even lower pay and possibly harsher working conditions for a given occupation. Specifically, noncollege workers appear to have diminishing chances at getting jobs at high-wage employers, as seen in a rise in the correlation of firm wage premiums with worker education and worker wage-fixed effects (the permanent wage component that persists across employers) both in the United States and Europe. David Card, Jörg Heining, and Patrick Kline, "Workplace Heterogeneity and the Rise of West German Wage Inequality," *Quarterly Journal of Economics* 128, no. 3 (2013): 967–1015; and David Autor et al., "The Fall of the Labor Share and the Rise of Superstar Firms," *Quarterly Journal of Economics* 135, no. 2 (2020): 645–709.

4. David H. Autor, Frank Levy, and Richard J. Murnane, "The Skill Content of Recent Technological Change: An Empirical Exploration," *Quarterly Journal of Economics* 118, no. 4 (2003): 1279–1333.

5. For an insightful analysis of job polarization along routine versus nonroutine task-intensive occupations, see Maarten Goos and Alan Manning, "Lousy and Lovely Jobs: The Rising Polarization of Work in Britain," *Review of Economics and Statistics* 89, no. 1 (February 2007): 118–133, https://www.jstor.org/stable/40043079.

NOTES TO CHAPTER 9   305

6. Juanita Constible, Nikole Blandon, and Bora Chang, "On the Frontlines: Climate Change Threatens the Health of America's Workers," Natural Resources Defense Council, July 28, 2020, https://www.nrdc.org/resources/frontlines-climate-change-threatens-health-americas -workers.

7. For more detailed discussion of the tension between compensating differentials and utility dispersion, see Dale Mortensen, *Wage Dispersion: Why Are Similar Workers Paid Differently?* (Cambridge, MA: MIT Press, 2003); Isaac Sorkin, "Ranking Firms Using Revealed Prefer- ence," *Quarterly Journal of Economics* 133, no. 3 (2018): 1331–1393.

8. R. Jisung Park and Paul Stainier, "The Distribution of Workplace Climate Amenities" (working paper, 2022).

9. Injury rates come from the Bureau of Labor Statistics, "Injuries, Illnesses, and Fatalities," accessed June 2022, https://www.bls.gov/iif/oshcfoi1.htm#2020.

10. Roughly 0.7 out of 100 full-time equivalent (FTE) workers employed in professional, scientific, and technical services are seriously hurt or fall ill on the job each year. For managers, the rate is 0.6 per 100 FTE workers.

11. R. Jisung Park, Nora Pankratz, and A. Patrick Behrer, "Temperature, Workplace Safety, and Labor Market Inequality" (IZA Discussion Paper 14560, 2021), https://docs.iza.org /dp14560.pdf.

12. Marcus Dillender, "Climate Change and Occupational Health: Are There Limits to Our Ability to Adapt?," *Journal of Human Resources* 56, no. 1 (2021): 184–224.

13. For instance, see Daniel S. Hamermesh, "Changing Inequality in Markets for Workplace Amenities," *Quarterly Journal of Economics* 114, no. 4 (1999): 1085–1123.

14. While we are unable to match injury risk to individual wages and must approximate it using average income at the level of residential zip code, we believe that doing so may understate the extent of inequality across income groups. If a gradient in injury rates by individuals' income exists and is downward sloping, and there are low-income individuals living in high-income zip codes, our estimates of the injury rate of high-income individuals may be biased upward. Con- versely, if there are high-income individuals living in low-income zip codes, the estimated injury rate of low-income individuals will be biased downward. Both of these biases imply that, were we able to estimate using individual rather than area income, the gradient across individuals would be steeper than the gradient we estimate here across zip codes.

15. Jacqueline M. Doremus, Irene Jacqz, and Sarah Johnston, "Sweating the Energy Bill: Extreme Weather, Poor Households, and the Energy Spending Gap," *Journal of Environmental Economics and Management* 112 (2022): 102609.

16. Jason DeParle, "Vast Expansion in Aid Kept Food Insecurity from Growing Last Year," *New York Times*, September 11, 2021, https://www.nytimes.com/2021/09/08/us/politics/vast -expansion-aid-food-insecurity.html.

17. Ten percent of households reported keeping their dwelling at an unhealthy or unsafe temperature for at least one month in 2015. But this masks enormous variation. The proportion, at 20 percent, is far higher for households in extreme poverty (20 percent for those making less than $20,000 per year; 14 percent for those making between $20,000 and $39,999). This com- pares to 3 percent for those making more than $100,000 per year. Over 20 percent said they reduced or went without basic necessities such as food or medicine to pay a home energy bill. Black and Hispanic households are more than twice as likely as whites to do this: 36 percent

versus 17 percent (4.9 out of 13.6 million households for Blacks versus 14 out of 82 million for whites). US Energy Information Administration, "Residential Energy Consumption Survey (RECS)," 2015, https://www.eia.gov/consumption/residential/data/2015/.

18. Our data comes from Southern California Edison, a major utility that serves over 15 million people in central, coastal, and southern California. The data include the account histories of 300,000 low-income households across 300 zip codes in the utility company's service area between January 2012 and August of 2017.

19. Alan Barreca, R. Jisung Park, and Paul Stainier, "High Temperatures and Electricity Disconnections for Low-Income Homes in California," *Nature Energy* 7, no. 11 (2022): 1052–1064.

20. For more detailed qualitative evidence regarding these and other hardships faced by low-income families, see Colleen Heflin, Andrew London, and Ellen K. Scott, "Mitigating Material Hardship: The Strategies Low-Income Families Employ to Reduce the Consequences of Poverty," *Sociological Inquiry* 81, no. 2 (May 2011): 223–246, https://doi.org/10.1111/j.1475-682X .2011.00369.x.

21. See Jayanta Bhattacharya et al., "Heat or Eat? Cold-Weather Shocks and Nutrition in Poor American Families," *American Journal of Public Health* 93, no. 7 (2003): 1149–1154; Timothy K. M. Beatty, Laura Blow, and Thomas F. Crossley, "Is There a 'Heat-or-Eat' Trade-Off in the UK?," *Journal of the Royal Statistical Society Series A: Statistics in Society* 177, no. 1 (January 2014): 281–294.

22. This is due primarily to the fact that the value of statistical life (VSL), which is the primary intellectual tool used to evaluate health benefits of policies, is a "revealed preference" metric, which depends both on willingness and ability to pay. The upshot is that poorer people are "revealed" to have a lower economic valuation of their health and safety, which means that the VSL in a poor country like India is mechanically lower than in a rich country like Ireland.

23. Solomon Hsiang et al., "Estimating Economic Damage from Climate Change in the United States," *Science* 356, no. 6345 (2017): 1362–1369.

Chapter Ten

1. United Nations Environment Programme, "As Impacts Accelerate, Adapting to Climate Change Must Become a Global Priority," November 3, 2022, https://www.unep.org/news-and -stories/press-release/impacts-accelerate-adapting-climate-change-must-become-global.

2. United Nations Environment Programme, "As Impacts Accelerate."

3. Ishan B. Nath, "The Food Problem and the Aggregate Productivity Consequences of Climate Change" (working paper 27297, National Bureau of Economic Research, 2020), https:// doi.org/10.3386/w27297.

4. Frances C. Moore et al., "New Science of Climate Change Impacts on Agriculture Implies Higher Social Cost of Carbon," *Nature Communications* 8, no. 1 (2017): 1607.

5. One important caveat is that such techniques are unable to tell us much about the role of future technical change or about the possibility of future policy regimes that look different from current and past regimes; we will revisit this momentarily. But to a first approximation, such analyses provide incredibly useful policy inputs from the perspective of understanding how effective adaptation could be.

NOTES TO CHAPTER 11   307

6. Tamma Carleton et al. "Valuing the Global Mortality Consequences of Climate Change Accounting for Adaptation Costs and Benefits," *Quarterly Journal of Economics* 137, no. 4 (2022): 2037–2105.

7. Nath, "Food Problem."

8. Robin Burgess et al., "Weather, Climate Change and Death in India" (working paper, University of Chicago, 2017), https://epic.uchicago.edu/wp-content/uploads/2019/07/Publication-9.pdf.

9. Burgess et al., "Weather, Climate Change and Death in India."

10. Dean Karlan et al., "Agricultural Decisions after Relaxing Credit and Risk Constraints," *Quarterly Journal of Economics* 129, no. 2 (2014): 597–652.

11. Gregory Lane, "Adapting to Floods with Guaranteed Credit: Evidence from Bangladesh" (working paper, Harris School of Public Policy, University of Chicago, 2022), https://www.gregoryvlane.com/pdfs/Lane_Floods.pdf.

12. Nath, "Food Problem."

13. Frances C. Moore, Uris Baldos, Thomas Hertel, and Delavane Diaz, "New Science of Climate Change Impacts on Agriculture Implies Higher Social Cost of Carbon," *Nature Communications* 8, no. 1 (2017): 1607.

14. Manisha Shah and Bryce Millett Steinberg, "Drought of Opportunities: Contemporaneous and Long-Term Impacts of Rainfall Shocks on Human Capital," *Journal of Political Economy* 125, no. 2 (2017): 527–561.

15. Shah and Steinberg, "Drought of Opportunities," 539–545.

16. See Harry Anthony Patrinos and George Psacharopoulos, "Returns to Education in Developing Countries," in *The Economics of Education*, ed. Steve Bradley and Colin Green (London: Academic Press, 2020), 53–64; and James J. Heckman, "Skill Formation and the Economics of Investing in Disadvantaged Children," *Science* 312, no. 5782 (2006): 1900–1902.

17. Teevrat Garg, Maulik Jagnani, and Vis Taraz, "Temperature and Human Capital in India," *Journal of the Association of Environmental and Resource Economists* 7, no. 6 (2020): 1113–1150.

18. Ayushi Narayan, "The Impact of Extreme Heat on Workplace Harassment and Discrimination," *Proceedings of the National Academy of Sciences* 119, no. 39 (2022): e2204076119.

19. Suresh Naidu, Yaw Nyarko, and Shing-Yi Wang, "Monopsony Power in Migrant Labor Markets: Evidence from the United Arab Emirates," *Journal of Political Economy* 124, no. 6 (2016): 1735–1792.

20. Matthieu Glachant, "Innovation in Climate Change Adaptation: Does It Reach Those Who Need It Most?," June 9, 2020, https://blogs.worldbank.org/climatechange/innovation-climate-change-adaptation-does-it-reach-those-who-need-it-most.

21. Joseph Stiglitz's book *Globalization and Its Discontents* (New York: Norton, 2002) provides one of many reminders.

Chapter Eleven

1. Damian Carrington, "Why the Guardian Is Changing the Language It Uses about the Environment," *The Guardian*, May 17, 2019, https://www.theguardian.com/environment/2019/may/17/why-the-guardian-is-changing-the-language-it-uses-about-the-environment; quoted in Guardian Staff, "The Climate Emergency Is Here. The Media Needs to Act Like It," *Guardian*,

April 12, 2021, https://www.theguardian.com/environment/2021/apr/12/covering-climate-now-guardian-climate-emergency#:~:text=Two%20years%20ago%2C%20the%20Guardian,appropriately%20urgent%20%E2%80%9Cclimate%20emergency%E2%80%9D.

2. Covering Climate Now, "Statement on the Climate Emergency," accessed June 2022, https://coveringclimatenow.org/projects/covering-climate-now-statement-on-the-climate-emergency/.

3. Ashlee Cunsolo et al., "Ecological Grief and Anxiety: The Start of a Healthy Response to Climate Change?," *The Lancet Planetary Health* 4, no. 7 (July 2020): e261–e263.

4. Caroline Hickman et al., "Climate Anxiety in Children and Young People and Their Beliefs about Government Responses to Climate Change: A Global Survey," *The Lancet Planetary Health* 5, no. 12 (2021): e863–e873.

5. David Wallace-Wells, "The Uninhabitable Earth," *New York*, July 10, 2017, https://nymag.com/intelligencer/2017/07/climate-change-earth-too-hot-for-humans.html; David Wallace-Wells, *The Uninhabitable Earth: Life after Warming* (New York: Tim Duggan Books, 2020). The book was named one of the best books of the year by *The New Yorker*, *The New York Times Book Review*, *Time*, NPR, *The Economist*, *The Paris Review*, the *Toronto Star*, and *GQ*.

6. Shannon Osaka, "The World's Most Ambitious Climate Goal Is Essentially Out of Reach," Grist, April 8, 2022, https://grist.org/climate/the-worlds-most-ambitious-climate-goal-is-essentially-out-of-reach/; and Brad Plummer and Nadja Popovich, "Why Half a Degree of Global Warming Is a Big Deal," *New York Times*, October 7, 2018, https://www.nytimes.com/interactive/2018/10/07/climate/ipcc-report-half-degree.html.

7. Most analyses of the potential melting of the Greenland Ice Sheet consider an equilibrium time frame of approximately 300 years to a few millennia. While global mean sea level during the mid-Pliocene warm period, when temperatures were 2–4°C warmer than preindustrial periods, was estimated to be up to twenty-five meters higher, the multimillennial time scales of these past climate and sea-level changes means that they are of limited value in projecting future sea level rise on scales relevant for most human policymaking. Similarly, while recent estimates suggest that the melting of the Greenland Ice Sheet could lead to up to 7.2 meters of eventual sea level rise, its contribution to sea level rise over the next century may be on the order of 5 to 33 centimeters. See Xavier Fettweis et al., "Estimating the Greenland Ice Sheet Surface Mass Balance Contribution to Future Sea Level Rise Using the Regional Atmospheric Climate Model MAR," *The Cryosphere* 7, no. 2 (2013): 469–489; and Andy Aschwanden et al., "Contribution of the Greenland Ice Sheet to Sea Level over the Next Millennium," *Science Advances* 5, no. 6 (2019): eaav9396.

8. Nicholas Stern and Joseph E. Stiglitz, "The Social Cost of Carbon, Risk, Distribution, Market Failures: An Alternative Approach" (working paper 28472, National Bureau of Economic Research, 2021).

9. The 2007 IPCC report uses the period from 1980 to 1999 as the baseline from which additional warming is calculated, whereas the latest 2022 IPCC report uses the preindustrial period (from 1850 to 1900). The latest report notes that this is done by adding 0.85°C to simulated changes relative to 1995–2014.

10. Brad Plumer and Nadja Popovich provide a highly accessible summary of these changes, though there is considerable disagreement about how to characterize model uncertainty, particularly regarding the higher-end warming scenarios. "Yes, There Has Been Progress on

Climate. No, It's Not Nearly Enough," *New York Times*, October 25, 2021, https://www.nytimes.com/interactive/2021/10/25/climate/world-climate-pledges-cop26.html.

11. David Wallace-Wells, "Beyond Catastrophe: A New Climate Reality Is Coming into View," *New York Times*, October 26, 2022, https://www.nytimes.com/interactive/2022/10/26/magazine/climate-change-warming-world.html.

12. Stanley Reed, "Giant Wind Farms Arise off Scotland, Easing the Pain of Oil's Decline," *New York Times*, November 27, 2022, https://www.nytimes.com/2022/11/27/business/scotland-wind-farms-offshore.html.

13. At the time of this writing, the United Kingdom and the EU member states have committed to cutting economy-wide greenhouse gas emissions by at least 68 percent and 55 percent by 2030, respectively (relative to 1990 levels). One of the major policy instruments that helped make this happen is the cap-and-trade program (now two cap-and-trade programs, with the United Kingdom's departure from the EU). As the cap on emissions tightens, there are fewer and fewer allowances available for fossil fuel emitting companies to purchase in order to offset their emissions, which means that for companies that invest in renewables like offshore wind, their profitability increases along with the stringency of the cap. As of late 2022, the price of emissions allowances on the United Kingdom's emissions trading system stood at around £70 pounds ($90), the implication being that, on the margin, firms that face a cost of mitigating emissions of less than £70 pounds per ton CO_2 would make money by investing in such mitigation technologies and practices.

14. Technically, it is a distribution of a forward-looking schedule of numbers, depending on a range of parameters, whose central tendency we often report as "the social cost of carbon."

15. McKinsey and Company, "The State of Internal Carbon Pricing," February 10, 2021, https://www.mckinsey.com/capabilities/strategy-and-corporate-finance/our-insights/the-state-of-internal-carbon-pricing.

16. Christiana Figueres, Yvo de Boer, and Michael Zammit Cutajar, "For 50 Years, Governments Have Failed to Act on Climate Change," *The Guardian*, June 2, 2022; Peter Gassmann, Casey Herman, and Colm Kelly, "Are You Ready for the ESG Revolution?," PwC Global, June 15, 2021, https://www.pwc.com/gx/en/issues/esg/esg-revolution.html; Apple newsroom, "Apple Commits to Be 100 Percent Carbon Neutral for Its Supply Chain and Products by 2030," July 21, 2020, https://www.apple.com/newsroom/2020/07/apple-commits-to-be-100-percent-carbon-neutral-for-its-supply-chain-and-products-by-2030/.

17. Milton Friedman, "A Friedman Doctrine—The Social Responsibility of Business Is to Increase Its Profits," *New York Times*, September 13, 1970, https://www.nytimes.com/1970/09/13/archives/a-friedman-doctrine-the-social-responsibility-of-business-is-to.html.

18. Quoted in Pascal-Emmanuel Gobry, "What Reagan's Greatest Economic Adviser Thought about Austerity," *Forbes*, June 5, 2013, https://www.forbes.com/sites/pascalemmanuelgobry/2013/06/05/milton-friedman-on-austerity/?sh=4752430d5628.

19. Milton Friedman, *Capitalism and Freedom* (Chicago: University of Chicago Press, 1962), 32.

20. See, for instance, "Economists' Statement on Carbon Dividends," *Wall Street Journal*, January 17, 2019, https://www.wsj.com/articles/economists-statement-on-carbon-dividends-11547682910.

21. Geoffrey Heal, *When Principles Pay: Corporate Social Responsibility and the Bottom Line* (New York: Columbia Business School Publishing, 2008), 36.

22. To be precise, it represents the change in the discounted value of economic welfare arising from a change in present emissions.

23. As quoted in Catherine Early, "The Most Important Number You've Never Heard Of," BBC, November 10, 2021, https://www.bbc.com/future/article/20211103-the-most-important-number-youve-never-heard-of.

24. Matthew Brown, "EXPLAINER: Can Climate Change Be Solved by Pricing Carbon?," AP News, April 22, 2022, https://apnews.com/article/climate-biden-business-billings-environment-0835d2e4f113ad1c2c26747c69d9e6bf.

25. Jessica Fan, Werner Rehm, and Giulia Siccardo, "The State of Internal Carbon Pricing," McKinsey & Company, February 10, 2021, https://www.mckinsey.com/capabilities/strategy-and-corporate-finance/our-insights/the-state-of-internal-carbon-pricing.

26. CDP, "Nearly Half of World's Biggest Companies Factoring Cost of Carbon into Business Plans," April 21, 2021, https://www.cdp.net/en/articles/media/nearly-half-of-worlds-biggest-companies-factoring-cost-of-carbon-into-business-plans.

27. Nordhaus notes that the vast majority of sectors in the economy are unaffected based largely on informed introspection: "Table 5 shows a sectoral breakdown of United States national income, where the economy is subdivided by the sectoral sensitivity to greenhouse warming. The most sensitive sectors are likely to be those, such as agriculture and forestry, in which output depends in a significant way upon climatic variables. At the other extreme are activities, such as cardiovascular surgery or microprocessor fabrication in 'clean rooms,' which are undertaken in carefully controlled environments that will not be directly affected by climate change. Our estimate is that approximately 3% of United States national output is produced in highly sensitive sectors, another 10% in moderately sensitive sectors, and about 87% in sectors that are negligibly affected by climate change." William D. Nordhaus, "To Slow or Not to Slow: The Economics of the Greenhouse Effect," *Economic Journal* 101, no. 407 (1991): 930.

28. William D. Nordhaus, "Rolling the 'DICE': An Optimal Transition Path for Controlling Greenhouse Gases," *Resource and Energy Economics* 15, no. 1 (1993): 27–50.

29. The EPA report relied heavily on the GIVE model per Kevin Rennert et al., "The Social Cost of Carbon: Advances in Long-Term Probabilistic Projections of Population, GDP, Emissions, and Discount Rates" (working paper no. 21-28, Resources for the Future, 2021), https://media.rff.org/documents/WP_21-28_V2.pdf, as well as damage functions from the Climate Impact Lab's data-driven spatial climate impact model and an aggregate damage function per Peter Howard and Thomas Sterner, "Few and Not So Far Between: A Meta-Analysis of Climate Damage Estimates," *Environmental and Resource Economics* 68, no. 1 (2017): 197–225. According to the report's table ES.1, the central EPA estimates for 2020 were $190 per ton of CO_2 for the social cost of greenhouse gases, $1,600 per ton of CH_4, and $54,000 per ton of N_2O, with all figures rounded to two significant digits. Using the GIVE model, Rennert et al. estimated the mean social cost of CO_2 to be $185 per ton CO_2 in 2020 dollars, with a 5–95 percent range of $44 to $413. Interagency Working Group on Social Cost of Greenhouse Gases, United States Government, "Technical Support Document: Social Cost of Carbon, Methane, and Nitrous Oxide Interim Estimates under Executive Order 13990," https://www.whitehouse.gov/wp-content/uploads/2021/02/TechnicalSupportDocument_SocialCostofCarbonMethaneNitrousOxide.pdf.

NOTES TO CHAPTER 11   311

30. The discount rate is especially critical in the context of climate mitigation policy because the benefits from climate policy accrue over multiple generations, while the costs are borne closer to the present. Consider the present value of an investment that pays out in the form of $10,000 in benefits realized 100 years in the future. Whether you value that future money more or less makes a big difference. At a 6 percent discount rate, the present value of the investment is a mere $29.47. At a 2 percent discount rate, the present value is $1,380.33, a difference of nearly forty-seven-fold.

There are, broadly speaking, two reasons for discounting future monetary flows. The first is what is referred to as the declining marginal value of consumption, which is based on the intuition that an additional dollar is worth more to a poor person relative to a rich one. An extension of this idea is that richer you will value an additional dollar less than poorer you; conversely, an added monthly expenditure of $1,000 is less of a hit to steady income-earning adult you than to debt-accumulating student you. The main relevance to climate change is that damages occurring in 2075 or 2100 may be more manageable, to the extent that societies are richer then compared to now.

The second is what is called the pure rate of time preference, which captures the idea that people value the future less than the present, irrespective of income. While as a descriptive phenomenon this impatience has been long recognized, whether it is normatively applicable to policymaking is debated, particularly in contexts like climate change where decisions made today have implications for multiple generations, many of whom have no say in the decision-making process. In other words, in settings where, to use Friedman's terminology, the "neighborhood effects" extend globally and across generations, it isn't clear that one generation's impatient market behaviors should form the basis for collective, transgenerational decision-making. One practical explanation for some nonzero but very small pure rate of time preference (e.g., 0.1 percent) might be the possibility of a major disaster that wipes out all humans at some future date, which would remove the value of any events that happen afterward. But most scholars agree that anything above and beyond this would be difficult to justify ethically, given the implications for how one values a life on the basis of the arbitrariness of when one is born.

31. Nicholas Stern, *The Economics of Climate Change: The Stern Review* (Cambridge: Cambridge University Press, 2007).

32. There are two distinct components of the social discount rate that are relevant here. The pure rate of time preference, and the marginal elasticity of consumption utility. The former captures societal impatience in the sense that it reflects a preference for consumption in the near as opposed to distant future; the latter reflects diminishing returns to additional units of consumption as income levels rise. The latter enters more as a technical matter in that it affects the social discount rate by way of projected economic growth. If economic growth is high, then we can expect incomes to be higher in the future, and the marginal utility associated with a given increment of income in this richer future to be valued less mechanically. The former is arguably more ethically determined. Stern and Nordhaus differed on both dimensions, but their opinions diverged most strongly with regard to the pure rate of time preference, which Stern argued should be very low (0.1 percent), and Nordhaus argued should be higher (3 percent).

33. William D. Nordhaus, "A Review of the *Stern Review on the Economics of Climate Change*," *Journal of Economic Literature* 45, no. 3 (September 2007): 686–702, https://doi.org/10.1257/jel

.45.3.686. As it turns out, there is at the time of this writing arguably much less disagreement over the "correct" discount rate than might be apparent to the casual observer. In brief, there is a growing recognition that, for most defensible ethical positions, the pure rate of time preference component of the social discount rate should be close to zero. There is also a growing consensus that the relevant risk-free interest rate to reference is likely not the average return on equities, as many, including Nordhaus, initially argued; in a world where future economic growth rates and their correlation with climate damages is uncertain, discount rates should be state-contingent.

34. Moritz A. Drupp et al., "Discounting Disentangled," *American Economic Journal: Economic Policy* 10, no. 4 (November 2018): 109–134, https://doi.org/10.1257/pol.20160240.

35. For an excellent review of these and other relevant developments in the discounting literature, see Antony Millner and Geoffrey Heal, "Choosing the Future: Markets, Ethics, and Rapprochement in Social Discounting," *Journal of Economic Literature* 61, no. 3 (2023): 1037–1087.

36. Kevin Rennert et al., "Comprehensive Evidence Implies a Higher Social Cost of CO_2," *Nature* 610 (2022): 687–692, https://doi.org/10.1038/s41586-022-05224-9.

37. Incorporating climate inequality into climate policy is a complex task that presents several challenges. Standard US government procedures typically involve conducting separate distributional analyses of policies, rather than integrating distributional concerns into a cost-benefit analysis. However, some experts argue that equity weighting should be incorporated into the calculation of the social cost of carbon (SCC) based on the fundamental economic concept of declining marginal value of consumption. This approach would involve calibrating equity weights from existing literature and applying them at the spatial resolution of damages. Despite the intellectual coherence of this approach, current US policy guidelines do not clearly allow for equity weighting within cost-benefit calculations. This means that incorporating equity concerns would represent a significant departure from standard US cost-benefit analysis, although it would not be legally indefensible. However, updates have been proposed to the US Office of Management and Budget's guidance on regulatory analysis (Circular A-4) that would potentially include more standardized language regarding distributional effects and how they should be incorporated in cost-benefit analyses of government regulations.

A practical alternative to equity weighting may be to simply display distributional impacts alongside a standalone SCC, demonstrating how damages vary by geography, income, race, and other factors both across and within countries. This approach could allow policymakers and the public to incorporate equity concerns into decision-making without imposing a particular value criterion chosen by experts working on the social cost of carbon.

One analysis by economists David Anthoff and Johannes Emerling showed that incorporating "inequality aversion"—which is essentially about how much we care about damages from climate change affecting poorer people, whether that's poorer people today or current generations relative to future generations (which will presumably be richer due to continued economic growth)—in integrated assessment models increases social cost of carbon estimates by a factor of 2 or 3, possibly more.

38. Goldman Sachs, "Carbonomics," accessed March 2023, https://www.goldmansachs.com/intelligence/topics/carbonomics.html.

NOTES TO CHAPTER 12   313

39. Ian Tiseo, "Greenhouse Gas Emissions per Capita in the European Union (EU-27) from 1990 to 2021," Statista, June 6, 2023, https://www.statista.com/statistics/986460/co2-emissions-per-cap-eu/.

40. Anmar Frangoul, "Tesla's Model 3 Was the UK's Second Most Popular New Car in 2021," CNBC, January 6, 2022, https://www.cnbc.com/2022/01/06/teslas-model-3-was-the-uks-second-most-popular-new-car-in-2021.html.

41. Bill Chappell, "9 in 10 Cars Now Being Sold in Norway Are Electric or Hybrid," NPR, October 8, 2021, https://www.npr.org/2021/10/08/1044330824/norway-electric-vehicle-car-sales-evs.

42. Mike Bird, Soumaya Keynes, and Alice Fulwood, "India's Moment," September 14, 2022, in *Money Talks*, podcast, 37:00, https://www.economist.com/indiapod.

Chapter Twelve

1. US Army Corp of Engineers, "NY & NJ Harbor & Tributaries Focus Area Feasibility Study (HATS)," New York District website, accessed July 2023, https://www.nan.usace.army.mil/Missions/Civil-Works/Projects-in-New-York/New-York-New-Jersey-Harbor-Tributaries-Focus-Area-Feasibility-Study/.

2. Jill Stoddard, "Our Common Future in a New Climate World: Interview with Robert Mendelsohn," Global Observatory, November 16, 2022, https://theglobalobservatory.org/2022/11/climate-interview-with-robert-mendelsohn/.

3. William Nordhaus, "Expert Opinion on Climate Change," *American Scientist* 82, no. 1 (1994): 45–51.

4. The White House, "Opportunities for Better Managing Weather Risk in the Changing Climate," in *Economic Report of the President* (Washington, DC: The White House, 2023), 273–304.

5. Anne Barnard, "The $119 Billion Sea Wall That Could Defend New York . . . or Not," *New York Times*, August 21, 2021, https://www.nytimes.com/2020/01/17/nyregion/the-119-billion-sea-wall-that-could-defend-new-york-or-not.html.

6. Danielle Ling, "Most CEOs, CFOs Say Addressing Climate Risk Is a Top Business Priority," PropertyCasualty360, July 16, 2020, https://www.propertycasualty360.com/2020/07/16/most-ceos-cfos-say-addressing-climate-risk-is-a-top-business-priority/.

7. Sandor Boyson et al., "How Exposed Is Your Supply Chain to Climate Risks?," *Harvard Business Review*, May 2, 2022, https://hbr.org/2022/05/how-exposed-is-your-supply-chain-to-climate-risks.

8. Jean-Noël Barrot and Julien Sauvagnat, "Input Specificity and the Propagation of Idiosyncratic Shocks in Production Networks," *Quarterly Journal of Economics* 131, no. 3 (2016): 1543–1592.

9. Nora Pankratz and Christoph Schiller, "Climate Change and Adaptation in Global Supply-Chain Networks," Proceedings of Paris December 2019 Finance Meeting EUROFIDAI-ESSEC (European Corporate Governance Institute–Finance working paper no. 775, 2021), https://papers.ssrn.com/sol3/papers.cfm?abstract_id=3475416.

10. Nora Pankratz, Rob Bauer, and Jeroen Derwall, "Climate Change, Firm Performance, and Investor Surprises," *Management Science*, March 21, 2023, https://doi.org/10.1287/mnsc.2023.4685.

314 NOTES TO CHAPTER 12

11. Ishan B. Nath, "The Food Problem and the Aggregate Productivity Consequences of Climate Change" (working paper 27297, National Bureau of Economic Research, 2020), https:// www.nber.org/papers/w27297.

12. This figure comes from Miyuki Hino and Marshall Burke, "The Effect of Information about Climate Risk on Property Values," *Proceedings of the National Academy of Sciences—PNAS* 118, no. 17 (2021): 1–9. The lower number comes from using a 7 percent discount rate; the higher number uses a 3 percent discount rate.

13. Miyuki Hino and Marshall Burke, "The Effect of Information about Climate Risk on Property Values," *Proceedings of the National Academy of Sciences—PNAS* 118, no. 17 (2021): 1–9.

14. Laura A. Bakkensen and Lint Barrage, "Going Underwater? Flood Risk Belief Heterogeneity and Coastal Home Price Dynamics," *Review of Financial Studies* 35, no. 8 (2022): 3666–3709.

15. The White House, "Opportunities for Better Managing Weather Risk."

16. Jeffrey G. Shrader, Laura Bakkensen, and Derek Lemoine, "Fatal Errors: The Mortality Value of Accurate Weather Forecasts" (working paper no. 31361, National Bureau of Economic Research, 2023), https://www.nber.org/papers/w31361; Mitch Downey, Nelson Lind, and Jeffrey G. Shrader, "Adjusting to Rain before It Falls," *Management Science*, February 10, 2023, https://pubsonline.informs.org/doi/abs/10.1287/mnsc.2023.4697.

17. Francis Annan and Wolfram Schlenker, "Federal Crop Insurance and the Disincentive to Adapt to Extreme Heat," *American Economic Review* 105, no. 5 (2015): 262–266.

18. Kevin Watkins et al., *Human Development Report 2007/2008* (New York: United Nations Development Programme, 2007), https://hdr.undp.org/system/files/documents/human -development-report-20072008-english.2008-english.

19. We conducted a nationwide survey of school air-conditioning penetration and find based on student and counselor reports that in many parts of the United States fewer than 50 percent of classrooms appear to be adequately air-conditioned. A building condition survey of New York City public schools found that fewer than 68 percent had fully functioning climate control systems without any defective parts. R. Jisung Park et al., "Heat and Learning," *American Economic Journal: Economic Policy* 12, no. 2 (2020): 306–339.

20. Past analyses using process-based models provide valuable insights into climate change's impact on food systems by detailing the natural processes affecting crop yields such as root depth, evapotranspiration, and light utilization, and are fine-tuned with experimental fields. However, these models have yet to incorporate data on the real decisions made by farmers. Instead, they typically assume that farmers aim to maximize yields under externally imposed constraints, which are usually determined by the modeler, such as "no adaptation" or "optimal varietal switching." Like many scenario-based analyses, these scenarios represent what is theoretically achievable but do not necessarily reflect the actual conditions faced by farmers in different socioeconomic contexts. These real-world conditions are influenced by factors like credit limitations, incomplete information, market failures, and various social and environmental circumstances. See Andrew Hultgren et al., "Estimating Global Impacts to Agriculture from Climate Change Accounting for Adaptation," October 6, 2022, https://papers.ssrn.com/sol3 /papers.cfm?abstract_id=4222020.

21. Ashwin Rode et al., "Estimating a Social Cost of Carbon for Global Energy Consumption," *Nature* 598, no. 7880 (2021): 308–314; Hultgren et al., "Estimating Global Impacts";

Maximilian Auffhammer, "Climate Adaptive Response Estimation: Short and Long Run Impacts of Climate Change on Residential Electricity and Natural Gas Consumption," *Journal of Environmental Economics and Management* 114 (2022): 102669.

22. Patrick Baylis and Judson Boomhower, "Moral Hazard, Wildfires, and the Economic Incidence of Natural Disasters" (working paper no. 26550, National Bureau of Economic Research, 2019), https://www.nber.org/system/files/working_papers/w26550/w26550.pdf.

23. Alan Barreca et al, "Adapting to Climate Change: The Remarkable Decline in the US Temperature–Mortality Relationship over the Twentieth Century," *Journal of Political Economy* 124, no. 1 (2016): 105–159.

24. Park et al., "Heat and Learning."

25. Jamie T. Mullins and Corey White, "Can Access to Health Care Mitigate the Effects of Temperature on Mortality?," *Journal of Public Economics* 191 (2020): 1–15.

26. Mullins and White used data from the universe of Medicare claims combined with local daily weather data, as in previous studies. However, an interesting twist is that they also incorporated information on when community health centers were established in a particular location. Relying on the quirks of the relevant health law that led to a staggered rollout of grants for such centers, they were able to show how one causal relationship is causally affected by another, that is, how the causal effect of heat on mortality is affected (arguably causally) by the availability of local community health centers. They found that having more community health centers per capita had a significant protective effect, with a 1 percent increase in the number of community health centers per capita leading to a 0.6 percent reduction in heat-related mortality. This effect is of a similar magnitude to the effect of home air-conditioning, and greater than the effect of the number of doctors per capita. Mullins and White, "Access to Health Care."

27. Alan Barreca, R. Jisung Park, and Paul Stainier, "High Temperatures and Electricity Disconnections for Low-Income Homes in California," *Nature Energy* 7, no. 11 (2022): 1052–1064.

28. David Autor et al., "An Evaluation of the Paycheck Protection Program Using Administrative Payroll Microdata," *Journal of Public Economics* 211 (2022): 104664.

29. Guanghua Chi et al., "Microestimates of Wealth for All Low-and Middle-Income Countries," *Proceedings of the National Academy of Sciences* 119, no. 3 (2022): e2113658119.

Conclusion

1. In technical terms, this is what is called *existence value*, a subset of the *intrinsic value* associated with biodiversity. I, and likely many others, would certainly be willing to pay some amount to make sure they stayed that way. More importantly, even this rudimentary exposure to environmental ecology imparted a sense that the biodiversity rich forests in which such species reside offer innumerable benefits to animals and humans alike, including such "ecosystem services" as air and water purification, crop pollination, and carbon sequestration.

2. Rudolf de Groot et al., "Global Estimates of the Value of Ecosystems and Their Services in Monetary Units," *Ecosystem Services* 1, no. 1 (July 2012): 50–61.

3. International Union for Conservation of Nature, "Coral Reefs and Climate Change," March 2021, https://www.iucn.org/resources/issues-brief/coral-reefs-and-climate-change#:~:text=Coral%20reefs%20harbour%20the%20highest,combined%20with%20growing%20local%20pressures.

4. Kenneth R. N. Anthony, "Coral Reefs under Climate Change and Ocean Acidification: Challenges and Opportunities for Management and Policy," *Annual Review of Environment and Resources* 41 (2016): 59–81, https://doi.org/10.1146/annurev-environ-110615-085610.

5. Joan A. Kleypas, John W. McManus, and Lamber A. B. Meñez, "Environmental Limits to Coral Reef Development: Where Do We Draw the Line?," *American Zoologist* 39, no. 1 (February 1999): 146–159; and Anthony, "Coral Reefs under Climate Change."

INDEX

Note: Page numbers in *italics* indicate figures.

Abdul-Jabbar, Kareem, 69, 291n19
Abu Dhabi, 100
Acemoglu, Daron, 157
adaptation apartheid, 262
Administrative Procedures Act (1946), 143
agriculture: banking and insurance access,
  201–202, 263; domestic agriculture, 202;
  farming decisions, 264; insurance for losses,
  259; market access, 202–203; mortality
  statistics and, 81, 99; poor countries and,
  197; subsistence, 150–151, 200; weather and,
  203–204, 310n27, 314n20
Ainge, Danny, 70
air-conditioning: adaptation strategies and,
  109, 125, 159, 242; automobiles and, 133–134;
  British Navy studies, 104; criminal justice
  and, 146; disadvantaged minorities and,
  86–87, 176, 190–193; homes and, 7, 82, 114,
  150, 256, 258, 269, 272, 315n26; hydrofluorocarbons and, 226; Indian urban areas
  and, 292n11; indoor arenas and, 67–69,
  296n20; investments in, 16; mortality
  statistics and, 269–271, 292n18; prisons
  and, 144–145; schools and students, 50–51,
  72, 75–76, 108, 113, 264, 269–270, 314n19;
  Singapore and, 5, 149; working conditions
  and, 183, 246–247, 249; workplace injuries,
  188–189; worldwide access, 100, 151
air pollution, 58–61, 289n17
Albouy, David, 171, 173
Amazon employment, 177–179, 183, 304n3

American Economic Journal, 123
American Time Use Survey, 300n27
Amnesty International, 126
anecdotes and vignettes, 20–21
Angrist, Josh, 88
Anthoff, David, 312n37
Anthony, Kenneth, 280
Apple, 223, 248
assisted evolution, 281
asylum seekers and refugees, 126–127, 132,
  140–141, 143, 297n2
attribution science, 146
Auffhammer, Max, 264
Augustulus, Romulus, 3
Australia: climate anxiety, 215; climate
  skeptics, 26, 285n16; Great Barrier Reef,
  277–280; mitigation action, 195; wildfires
  and, 57, 268
Autor, David, 157, 181, 272

Bakkensen, Laura, 254, 258
Baltimore, Maryland, 174
Bangkok, Thailand, 100, 149, 158, *161*, 162,
  164, *165*, 168–169, 198
Ban Ki Moon, 127
Barrage, Lint, 254
Barreca, Alan, 191, 269, 271
Barrot, Jean-Noël, 248–249
Bauer, Rob, 249
Baylis, Patrick, 136–138, 266, 299n23
Behrer, Patrick, 50, 97, 144, 187

"Beyond Catastrophe" (Wallace-Wells), 219
big data and causality, 17, 73, 78, 116, 120, 138, 228, 272–273
Bill and Melinda Gates Foundation, 206–207
Bird, Larry, 69, 76, 99
Bloomberg, Michael, 75
Bolivia, 170
Bolotnyy, Valentin, 144
Boomhower, Judson, 266
Boston, 34, 57, 69–70, 157, 286n6
Boston Celtics, 69–70, 76
Boston Globe, 69, 76
Brazil, 50–51, 129, 170, 199, 209, 215
Breen, Mike, 68
British Navy studies, 103–108, 111, 294n7
British Petroleum (BP), 222
Bureau of Labor Statistics, 81, 187
Burgess, Robin, 201
Burke, Doris, 67
Burke, Marshall, 60, 115, 166, 254, 257

California: Camp Fire, 53; electricity disconnection, 192, 271; Fremont automobile factories, 117; heat wave, 84; lower average income and, 170–171; lower average income and temperature, 171–172; Paradise and Concow Fires, 53; Skirball fire, 39, 53, 55, 220; temperature and injuries, 97; wildfires, 39, 53–54, 56, 158, 266–268; workers' compensation claims, 187
California Division of Occupational Health and Safety (Cal/OSHA), 97
Canada: cold temperature and mortality, 169; food export economics, 203; greenhouse gas emissions, 151; heat exposure, 82; hot and cold marginal effects, 90, 100, 119–120; inequality in, 156; mitigation action, 195; wealth and climate, 118, 121; wildfires and smoke, 57–58
Capitalism and Freedom (Friedman), 223
carbon emission costs: discount rates, 225, 228, 230–231, 234–235; government role, 224; slow burn costs, 220; social cost, 121,

222, 225–235, 237, 287n2, 309n14, 310n22, 310n29, 311n30, 312n37
Card, David, 88
Carleton, Tamma, 139
Carney, Mark, 252
Cattaneo, Cristina, 141
Caulkins, Martha, 171
causality and correlation, 110, 295nn14–15
Centers for Disease Control and Prevention (CDC), 82–83, 90, 95–96, 294n30
Chen, Daniel, 144
Chen, Yuyu, 289n15
Chetty, Raj, 45, 157
Chicago, 42, 57, 199
Chicago heat wave (1995), 84, 87
China, 8–9, 25, 58, 114, 119, 124, 154–156, 209, 248
"Civilization and Climate" (Huntington), 103
Clay, Karen, 269
Clean Air Act, 58–59
climate adaptation: asymmetric information and, 257–258; credit constraints and, 258–259; engineering estimates and, 263–264; government's role in, 243, 245, 265–266, 271–272, 275–276; hotter climates and, 199; inequality and, 19–20; labor markets and, 260; moral hazards and, 20, 259; partial progress, 219–220; philanthropy and, 206, 208, 210–211; poorer countries and, 167, 198; private market's role, 250, 256, 274–275; questions of how and at what cost, 243; rich countries, financing pledge, 196–197; sea walls, 241–242; vulnerability and resilience, 159–160, 200, 213; vulnerable populations and, 261–263; weather forecasts and, 5, 258
climate anxiety, 215–217
climate catastrophe, 1–2, 4
climate change: moderate, noncatastrophic warming, 9; speed of, 7
climate change policies: economics of, 4, 222–225, 278; energy industry, 221–222

INDEX

Climate Framework for Uncertainty, Negotiation and Distribution (FUND) model, 227–229

climate hazards: corporate vulnerabilities, 250; exposure variations, 160–162, 190; hazard-prone area subsidies, 243; housing market and, 10; labor exposure and, 183, 185, 206; local circumstances of, 146; subsistence farmers and, 200, 210; understanding causes of, 92; vulnerability to, 159

Climate Impact Lab, 92–93, 95, 169, 198, 229, 264, 293n27, 310n29

climate justice, 9–10, 151, 175, 209, 214

climate mitigation: binary decision or a spectrum, 62; cap-and-trade programs, 309n13; discount rate and, 311n30, 311n32, 312n33; ecology and, 280–281; economic trade-offs and, 11, 15; government's role in, 282; greenhouse gas, 213, 222, 235; housing data and wildfires, 266; influences on, 16–17; marginal abatement and, 236; moral hazards and, 259; philanthropy and, 206, 208; poverty and, 194–195; reduction costs, 5, 237, 240; social costs, 233; technological advances, 218; US climate legislation, 8

climate risk: existential crisis, 11, 13–15, 55, 279; financial markets and, 12, 251–252; flood plains, 254–255; Global North and Global South, 208; government insurance and, 255; housing market and, 252–256, 266–268; physical climate risk, 12, 243–245, 250–253; tipping points, 14–15, 216–217, 233, 240; transition risk, 244–245, 252; worldwide impact, 15–16

community health clinics (CHCs), 270–271, 275, 315n26

coral reefs, 279–282

corporations and climate change: electricity grids and outages, 247–248; risk analytics businesses, 250–251, 275; supply-chain risks, 11–12, 118, 248; vulnerability and, 246–249

credibility revolution, 17, 88

creeping stressor, 280–281

criminal justice: expenditures, 128; rule of law and temperature, 143–144; sentencing and temperature, 144–145

Cropper, Maureen, 171

Cuba, 123

Dallas, Texas, 174

Darfur conflict, 127

Dell, Melissa, 123–124, 297n33

DeLong, Bradford, 155

Democratic Republic of the Congo, 155

Denver, Colorado, 174

Derwall, Jeroen, 249

Deschênes, Olivier, 87, 89, 91, 201, 269

Detroit, Michigan, 117

developing countries: adaptation and resilience in, 201; climate change costs, 9, 11; corporations in, 247; data collection issues, 273; financial opportunities, 204–205; greenhouse gas reductions, 239; heat-related mortality, 91–93, 167, 269; international assistance, 211, 243; labor market inequality, 181; market access, 202; poverty and climate needs, 197

Dietz, Simon, 233

Dillender, Marcus, 187

Djokovic, Novak, 115

Dominican Republic, 122–123

Donaldson, Dave, 201

Doremus, Jacqueline, 191

dose–response relationships, 85, 198–199, 264

Duncan, Tim, 67–68

dynamic integrated climate-economy (DICE) model, 120, 227–228

Ebenstein, Avraham, 113

economic inequality, 8, 36, 92, 157, 159, 181, 193, 275, 303n16

The Economist, 142

Egypt, 104, 203

El Salvador, 127, 129, 298n11

Emerling, Johannes, 312n37

320 INDEX

emotional affect and temperature, 136–138, 299n22

energy use, 191, 229, 264–265

environmental, social, and corporate governance (ESG) initiative, 221–222

Environmental Protection Agency (EPA), 56, 287n2, 310n29

equity weighting, 168, 232, 312n37

European heat wave (2003), 292n18

European Union (EU), 128, 140, 195, 221, 238–239, 300n30, 309n13

Evans, Hugh, 69

existence value, 315n1

exposure, definition of, 158

externalities, negative, 30, 225, 237, 261

externalities, positive, 237, 260–261, 266

Federal National Flood Insurance Program, 243

Fein, Steven, 135

Fink, Larry, 244

Forbes, 154

France: air-conditioning in, 100; climate anxiety, 215; European heat wave (2003), 292n18; heat wave mortality, 292n18; national disaster insurance, 243; per capita income, 153; poverty within, 209; school day closures, 51; wildfire risk, 268

Friedman, Milton, 223–224, 311n30

Garg, Teevrat, 205

Gates, Bill, 206, 208

Georgetown, Kentucky, 117

Gerarden, Todd, 239

Germany: cold and mortality, 90, 169; effect of heat in, 239; European heat wave (2003), 292n18; greenhouse gas contributions, 156; gun violence and homicide, 129; homicide rate, 298n12; hot and cold marginal effects, 100; poverty within, 209; schools and students, 50; solar subsidies, 239

Ghana, 100, 155, 169, 201, 232, 239, 302n3

Gini index, 156

GIVE model, 310n29

Glasgow Pledges (2021), 283n7

global climate justice, 197–198, 200

Global North and Global South, 9, 151, 167, 197, 209

Goldman Sachs Carbonomics platform, 236

Goodman, Josh, 48, 50, 114, 269

Gould, Ingrid, 298n7

Graf, Walter, 171

Graff-Zivin, Joshua, 60, 114

Graham, Benjamin, 251

granularity, 59, 115, 120, 124, 131, 195

Greece, 57, 153, 188n7

greenhouse gas emissions: benefits of mitigation, 7–8, 240; continued emissions cost, 6, 234–237; coral reefs and, 280; damage inequality, 194; economic costs, 279; emission reduction pledges, 309n13; energy transition timeframe, 238; equatorial countries and, 156; flow and stock of, 238, 311n30; high-income world and, 6, 151; natural variability and, 146; reduction methods, 213, 222; social cost, 225–226; United Nations report on, 283n8

Greenland ice sheet, 31, 216, 285n13, 308n7

Greenstone, Michael, 87, 89, 91, 201, 226, 269

Guanghua Chi, 273

The Guardian, 215, 222

Guterres, António, 196–197

Half the Sky (Kristof and WuDunn), 25–26

hazard, definition of, 158

Heal, Geoffrey, 225

Heilmann, Kilian, 130

Hendren, Nathan, 45, 157

hidden costs of climate change, 32–33

Hino, Miyuki, 254, 257

Hochul, Kathy, 58

Hornsea 2 wind farm, 220–221

Hsiang, Solomon, 122, 124, 166

human capital: air quality and learning, 61; definition of, 35; effect of disasters on, 45–47, 286nn5–6; heat and learning,

48–52, 71–75, 101, 290n13; monetary value of schooling, 44–45, 290n13; prenatal pollution exposure and, 287n15; student learning losses, 38, 42, 43, 44

human migration: intranational migration, 141; rising temperature effect, 141–143; slow-onset events and, 142; survival migration or profitable investment migration, 300n33

Huntington, Samuel, 103

Hurricane Katrina, 37–39, 47, 286n6, 286n9

Hurricane Laura, 123

Hurricane Rita, 123

Hurricane Sandy, 34–35, 37, 39, 47, 241, 246

Hurwitz, Mike, 114, 269

Husted, Lucas, 39

Imbens, Guido, 88

India: additional days over 32.2°C, 7; agriculture and rain, 203–204; air-conditioning in, 292n11; air pollution and mortality, 59; banking and, 201; British Navy studies, 104; climate anxiety, 215; criminal justice and heat, 144; greenhouse gas emissions, 156; heat and productivity, 118–119, 124, 199; mortality statistics and climate change, 16, 82, 91–92, 100, 164, 269; New Delhi and Kolkata, 292n11; poverty within, 209; rural banking and, 201; school performance, 50, 205; solar and wind implementation, 239; supply-chain issues, 247–249

inequality aversion, 312n37

Inflation Reduction Act (2022), 239, 283n8

informal workers, 205–206, 260

Institute for Economics and Peace, 127

integrated assessment model (IAM), 227–229, 232

Interagency Working Group on the Social Cost of Greenhouse Gases, 121

Intergovernmental Panel on Climate Change (IPCC), 21, 159, 218, 223, 308n9

Internal Revenue Service (IRS), 45, 157, 287n15

International Labour Organization (ILO), 205, 294n1

International Union for Conservation of Nature, 279

Israel, 61, 113, 124

Italy, 153, 156, 172, 292n18

Jackson, Mark, 68

Jacqz, Irene, 191

Jagnani, Maulik, 205

James, LeBron, 67–68, 71–72, 99, 106, 115

Japan, 90, 129, 156, 264, 298n12

Johnson, Magic, 69–70, 76–77, 106, 115

Johnston, Sarah, 191

Jones, Benjamin F., 123–124, 297n33

Jupiter Intel, 250

Kabore, Phillipe, 118

Kahn, Matthew, 130

Kahneman, Daniel, 27, 29, 31

Kansas City, Missouri, 84

Karlan, Dean, 201

Kellogg, Ryan, 171

Kenrick, Douglas, 132

Keynes, John Maynard, 155

Killea, Steve, 127

Kim Beom-su, 154

Kline, Patrick, 157

Klinenberg, Eric, 84

Knightian uncertainty, 11–12, 14, 217

Kristof, Nick, 25

labor market inequality: adaptation and, 260; automation and, 181–182; within countries, 156–158; education and, 180, 185–186, 304n3; environmental hazards and, 10, 180–181; hypotheses and, 181; outdoor work and workers, 99–100, 294n1; specialization, 180; workplace hazards and, 187, 305n10, 305n14

Larrick, Richard, 135

Lavy, Victor, 113

Lee Kwan Yew, 149

Lemoine, Derek, 258
Levy, Frank, 181
life expectancy, 59, 100, 154, 160, 289n15, 303n22
Lomborg, Bjorn, 12
Los Angeles, 1, 54–55, 57, 146
Los Angeles Lakers, 69–70, 76
Los Angeles Police Department (LAPD), 130–131
Los Angeles Times, 77
Luxembourg, 128

MacFarlane, Steven, 132
Mackworth, N. H., 104, 106, 108, 111
Malaysia, 129
Maldives, 208
Maxwell, Cedric, 69–70
McKinsey & Company, 222, 250
mental health and temperature, 138–139, 300n27
Mexico, 51, 101, 126, 129, 140, 170
Miami, Florida, 174
Miami Heat, 67–68, 72
Microsoft, 226
Miguel, Edward, 166
Mongolia, 102
Monroy, Ruth, 127
Montesquieu, 103
Moore, Frances, 203
mortality statistics and climate change:
African climate and, 93–94; banking and insurance access, 201–202; colder climates and, 90, 168–169; death certificates database, 87–89, 293nn20–22; elderly and milder heat events, 293n24; forward mortality displacement (harvesting), 85, 91, 94–95; heat-related mortality, 92–97, 198–199, 269, 315n26; heat wave case study approach, 84; less extreme heat and, 90–91; medical classification approach, 82–83; omitted variable bias, 88; traffic accidents, 79; undercounts, 83–84, 292n12; worldwide impact, 92–93, 169, 198

Mount Vesuvius, 1, 18
Mukherjee, Anita, 145
Mullins, Jamie, 138–139, 270, 300n27, 315n26
Murnane, Richard, 181

Nadal, Rafael, 115
Naidu, Suresh, 206
Namibia, 129
Nasheed, Mohamad, 208
Nath, Ishan, 199, 202
National Centers for Environmental Information (NCEI), 37, 286n5
National Climate Assessment, 21
National College Entrance Examination (NCEE), 114
National Football League (NFL), 135
nationalist dilemma of climate change, 31
national security and climate change, 140
Neidell, Matt, 60
the Netherlands, 82, 102, 110, 128, 239
net present value (NPV), 45
New Orleans, Louisiana. *See* Hurricane Katrina
New York City: New York Regents exams, 71–72, 73, 73–75, 109–110, *112*; reduction in very cold days, 168; smoke pollution, 57–58; temperature and humidity, 162; test scores and high temperature, 44–45, 113, 115, 124. *See also* Hurricane Sandy
New York sea wall, 241, 243–244, 255
New York Times, 75, 174, 216
New York Times/Siena College poll, 27
New Zealand, 156, 243
Nigeria, 102, 104, 110, 121, 203, 215
Nordhaus, William, 31, 227–228, 230, 310n27, 311n32, 312n33
Norfolk, Virginia, 117
North Korea, 102
Norway, 90, 121, 169, 239
Novartis, 226
Nuño-Sanchez, Jaime, 80–81, 83, 291n5
Nyarko, Yaw, 206

INDEX

O*NET, 184

occupational exposure, 183–186

Office Space (film), 65

Oklahoma tornadoes, 39

Olken, Benjamin A., 123–124, 297n33

omitted variable bias, 88, 102, 107–110, 129

Opper, Isaac, 39

Organisation for Economic Co-operation and Development (OECD), 151, 154, 156

Osei, Robert, 201

Osei-Akoto, Isaac, 201

Panama, 170

Pankratz, Nora, 97, 187, 248–249

Paris Agreement on climate change, 280, 283n7

Park, R. Jisung: Australian experience, 277; Los Angeles and smoke, 53–54; natural disasters and economics, 39; poverty and hot days, 191, 271; tests, learning and hot temperatures, 48, 50; worker exposure to climate hazards, 185, 187; workers' compensation claims, 97

Parker, Tony, 67

Paycheck Protection Program, 272

Peng Zhang, 114

Peri, Giovanni, 141

Peru, 170

Philadelphia, 34, 84

philanthropy, 206–208

the Philippines, 215

Phoenix, Arizona, 132–133

physical capital, 35–36, 39, 45–47, 57

Pinker, Stephen, 62, 126

poor countries: adaptation, 167; adaptation to heat, 199; air-conditioning in, 151; climate change and crop yields, 197; climate change and economic growth, 121, 124, 297n33; climate change consequences, 9, 209; domestic agriculture, 202; health and, 198, 306n22; targeting expenditures, 272; violent crime in, 146; vulnerability of, 10. *See also* agriculture

Popovich, Greg, 67–68

Portugal, 57, 158, 288n7, 292n18

poverty: adaptation policies for, 195, 273–274; adaptive behaviors and, 109; banking and, 201; climate damage and, 209–210; climate mitigation policies and, 194–195; damage to society, 193–194, 306n22; energy use and, 265, 271; hotter climates and, 130, 155; maximum temperatures and, 191–193, 305n17, 306n18; racism and, 175; rich countries, financing pledge, 196; violent crime and, 145–146; vulnerability to high temperature, 200

prisons and air-conditioning, 145–146

Proceedings of the National Academy of Sciences, 122

Providence, Rhode Island, 34

Puerto Rico, 123

PwC Global, 222–223

Qatar, 102

Qi Zhao, 95

Qu Tang, 114

Ranson, Matthew, 130

Refugee Act (1980), 298n2

Reifman, Alan, 135

resilience, 6, 159–160, 201, 211, 260, 262, 266, 281

Restrepo, Pascual, 157

Richmond, Virginia, 174

Risq, 250

Rivers, Nicholas, 118

road rage, 132–134, 299n17

Rockoff, Jonah, 45

Roman Empire, 2–3, 18, 383n5

Roth, Sefi, 113

Russia, 151, 221

Ryan, Bob, 69, 76

Sacerdote, Bruce, 286n6

Saez, Emmanuel, 157

San Antonio, Texas, 67–68

San Antonio Spurs, 67–68, 72

Sanders, Nicholas J., 145, 287n15

Saudi Arabia, 102

Sauvagnat, Julien, 248–249

Schiller, Christoph, 248

school closures and education: government aid and, 48; Hispanic and Black students, 50; Hurricane Katrina and, 37–38, 286n9; student learning losses, 42, 43, 44; Superstorm Sandy and, 34–35

Schwartz, Amy Ellen, 298n7

sea level rise, 31, 208–209, 216, 232, 246, 263, 285nn13–14, 308n7

Sen, Amartya, 25

Shah, Manishah, 203–204

Shapiro, Joseph, 269

Sharapova, Maria, 115

Sharkey, Patrick, 298n7

Shrader, Jeffrey, 258

Singapore, 102, 148–150, 158, 198

Sinha, Paramita, 171

Smith, Adam, 223

Smith, Jonathan, 48, 114, 269

South Africa, 129

South America, 100

South Asia, 51, 139

Southern California Edison, 306n18

South Korea: education and, 182–183; greenhouse gas emissions, 151; gun violence and homicide, 129; health insurance law, 301n1; high-speed rail, 153, 301n2; homicide rate, 298n12; inequality and, 156; mitigation action, 195; North Korea and economics, 102; per capita income, 302n5, 302n8; poverty within, 209; schools and students, 50–51; standard of living, 153–154

Spain, 96, 158, 243, 292n18

Special Flood Hazard Areas (SFHAs), 244

Stainier, Paul, 185, 191, 271

statistical or intuitive thinking, 25, 27–29, 31–32, 284n9, 285nn15–16

Stern, Nicolas, 230, 311n32

Sternberg, Bryce, 203–204

Stern Review on the Economics of Climate Change, 230

St. Louis, Missouri, 84

Sub-Saharan Africa, 51, 93

suicides, 5, 138–139, 155, 300n27

surveys, 26–27

Sweden, 90, 100

Syrian refugee crisis (2015), 127

Taiwan, 155

Taraz, Vis, 205

temperature and human health: brain and temperature stress, 107; elderly and cold avoidance, 173–174; heat exposure statistics, 100; heat illness and workers, 65–66, 80–81, 96–98, 291n5; heat waves and, 84–85, 89; poor countries and, 198, 306n5. *See also* mortality statistics and climate change

temperature and human performance: aggressive penalties in football, 135; air-conditioning and professional basketball, 67–70, 76–77, 291n19; batters hit by a ball, 135; cognitive activities and, 77–78, 103–107; colder temperatures and, 90; disadvantaged minorities or wealthier groups, 86–87, 190, 292n18; effect of air-conditioning, 113–114; extrapolation to other sports, 296n20; helicopter accidents and pilot error, 113; professional tennis matches, 115–116; serious injuries and, 4; student performance and school success, 71–75, 101, 109–114, 205, 290n13, 295n18; workplace injuries, 187–189; worldwide impact, 78, 82, 92, 292n11

temperature and human response, 210

temperature and per capita income, 100–103

temperature and productivity: adaptation investments, 119–120; adaptation strategies, 19, 125, 199, 296nn24–25; Canadian manufacturing study, 118–120; Chinese manufacturing study, 119; GDP growth and, 122–124, 297n33; India manufacturing study, 118–119; indoor effects, 120–121; overtime and, 296n22; US automobile manufacturing and, 117–119 temperature projection statistics, 7, 162–163, 164, 166, 172–173, 174, 218–219

Thailand, 7, 51, 82, 100, 150, 156, 162, 198, 248

Thames River Barrier, 241, 255

Thinking, Fast and Slow (Kahneman), 27

Thunberg, Greta, 12–13, 55, 62

time preference, 231, 311n30

Truman, Harry, 20

Turkey, 51

Tuskegee experiments, 108

Tutu, Desmond, 262

Tversky, Amos, 27, 29, 31

Udry, Christopher, 201

UNESCO World Heritage Convention, 279

"The Uninhabitable Earth" (Wallace-Wells), 216, 219

United Arab Emirates, 100, 206

United Kingdom: cold and mortality, 90, 169; emission reduction pledge, 309n13; greenhouse gas contributions, 156; gun violence and homicide, 129; health and hotter temperatures, 82; homicide rate, 298n12; hot and cold marginal effects, 100; poverty and neighborhoods, 172; public safety expenditures, 128; UK Climate Change Committee, 21

United Nations Convention relating to the Status of Refugees (1951), 297n2

United Nations Office on Drugs and Crime, 129

United Nations report on future adaptation needs, 196

United States: climate anxiety, 215; climate legislation, 8; emissions reductions, 239; greenhouse gas contributions, 156; heat-related mortality variations, 269; inequality and upward mobility, 157; life expectancy, 303n22; lower average income and temperature, 174–175, 177, 303n27; poverty within, 209; productivity and heat in, 199; redlining, 174–176; southern states and productivity, 119; standard of living, 154, 156, 302nn13–14; wildfire hazards in, 158

US Agency for International Development, 273

US Army Corps of Engineers, 241

US Department of Defense (DoD), 140

US National Ambient Air Quality Standards, 60–61

Uzbekistan, 102

value of statistical life (VSL), 306n22

Venezuela, 78, 170, 303n16

Vietnam, 247–248, 297n2

violent crime and high temperature, 128–132, 137, 139–140, 145–146, 298n7, 298n11, 299n15, 299n23

Vivid Economics, 250

vulnerability, 10, 18–19, 158–160, 190, 193–195, 205

Wallace-Wells, David, 216, 219

Wang, Shing-Yi, 206

Washington, DC, 34, 57, 165, 170

Wayne County, Michigan, 88

Weitzman, Martin, 30

When Principles Pay (Heal), 225

White, Corey, 138–139, 270, 300n27, 315n26

White House Council of Economic Advisers, 255

wildfires and smoke: Air Quality Index (AQI), 56, 61; climate change causation, 57, 288n7; construction regulations and insurance, 265–268; hidden health costs, 59–62, 220, 289n15, 289n17, 289n23; premature deaths from, 60; smoke pollution, 57–60. *See also* California; Canada

wildland-urban interface (WUI), 57

Wolff, Hendrik, 171

women and girls, killing of, 25–26

World Bank, 302n3, 302n5

Wu-Dunn, Sheryl, 25

X (formerly Twitter), 136

Yale Project on Climate Opinion, 26, 284n6

Yingquan Song, 114

A NOTE ON THE TYPE

This book has been composed in Arno, an Old-style serif typeface in the classic Venetian tradition, designed by Robert Slimbach at Adobe.

GPSR Authorized Representative: Easy Access System Europe - Mustamäe tee 50, 10621 Tallinn, Estonia, gpsr.requests@easproject.com

www.ingramcontent.com/pod-product-compliance
Lightning Source LLC
Jackson TN
JSHW020434140625
86093JS00003B/4